新编科技英语阅读教程

（上册）

主　编◎范莹芳　杨秀娟
副主编◎苑婷婷

哈尔滨工程大学出版社
Harbin Engineering University Press

内容简介

本书选材科学合理、内容丰富、难易兼顾，涵盖人工智能、神经技术、生态科学、信息安全、材料科学、海洋工程、核能应用、基因工程、空间开发、科学未解之谜等科学领域话题。通过科技英语语篇阅读，培养学生的英语水平、综合能力、思辨能力、家国情怀与国际视野。全书共分10个单元，每单元精选同一话题下多篇观点不同、相互补充或对立的文章设置课程内容与练习，引导学生多视角、全面、辩证地理解分析，进而提高语言能力和思辨能力，以更好地适应新时代发展对人才英语水平与思辨能力的要求。

本书主要面向理工科院校英语专业本科生、理工科院校非英语专业高年级本科生及研究生，也可供英语爱好者作为英语课外读物使用。

图书在版编目（CIP）数据

新编科技英语阅读教程. 上册 / 范莹芳，杨秀娟主编. — 哈尔滨：哈尔滨工程大学出版社，2020.8
ISBN 978-7-5661-2731-0

Ⅰ. ①新… Ⅱ. ①范… ②杨… Ⅲ. ①科学技术－英语－阅读教学－高等学校－教材 Ⅳ. ①N43

中国版本图书馆CIP数据核字(2020)第133323号

选题策划 雷 霞
责任编辑 张忠远 刘海霞
封面设计 张 骏

出版发行 哈尔滨工程大学出版社
社　　址 哈尔滨市南岗区南通大街145号
邮政编码 150001
发行电话 0451-82519328
传　　真 0451-82519699
经　　销 新华书店
印　　刷 哈尔滨市石桥印务有限公司
开　　本 787 mm×1 092 mm 1/16
印　　张 8.25
字　　数 227千字
版　　次 2020年8月第1版
印　　次 2020年8月第1次印刷
定　　价 27.00元
http://www.hrbeupress.com
E-mail:heupress@hrbeu.edu.cn

前　　言

本教材在内容设置上，通过科技英语语篇阅读，培养学生的英语水平、综合能力、思辨能力、家国情怀与国际视野，夯实语言功底、熟悉学科知识、培养综合能力、训练创新思维，引导学生立足新时代。

依据全国教育大会提出的培养学生综合能力、创新思维要求，结合《普通高等学校本科专业类教学质量国家标准》对英语专业学生思辨能力培养的要求，本教材力图在夯实学生语言基本功的同时，突破传统科技英语阅读教材集中关注语言能力训练的局限，适应新时代发展对学生能力与素养的需求，设置多样的内容形式，使学生在科技英语语篇阅读中训练分析、判断、评价、质疑、阐释等综合能力，提高思辨能力。同时，本教材拟避免传统科技英语阅读教材常见不足，结合国家对教育相关要求，突出课程思政元素，在材料选取上兼顾我国近年来在科技领域的前沿成果，增强学生四个自信，引领学生领会“科技是第一生产力，创新是引领发展的第一动力”的思想，增强“科技强国”意识与信念，以实现新时代教育应有之意。

本教材选材上努力做到科学合理、内容丰富、难易兼顾。选材涵盖人工智能、神经技术、生态科学、信息安全、材料科学、海洋工程、核能应用、基因工程、空间开发、科学未解之谜等科学领域话题。全书共分 10 个单元，每单元精选同一话题下多篇观点不同、相互补充或对立的文章设置内容与练习，引导学生多视角、全面、辩证地理解分析，进而提高语言能力和思辨能力，以更好地适应新时代发展对人才英语水平与思辨能力的要求。各单元课文 A 及课文 B 用于精读，课后设置丰富习题供学生进行思辨训练，包括思维导图、大纲填写、摘要撰写、句子释义、简答、观点评析、判断、翻译等多种形式。各单元附有一篇该主题领域补充阅读文章，供学生开展深入学习与训练。三篇课文配合使用能为课堂教学和学生课后学习与提高提供充足的内容，能够为学生思辨能力的提高提供较好的保障。

由于编者水平有限，本书存在诸多不足之处，恳请各位读者批评指正。

编　者

2019 年 10 月

CONTENTS

Unit One Artificial Intelligence

Critical Reading

Text A

Why Does Artificial Intelligence Scare Us So Much?

1. When people see machines that respond like humans, or computers that perform feats of strategy and cognition mimicking human ingenuity, they sometimes joke about a future in which humanity will need to accept robot overlords.
2. But buried in the joke is a seed of unease. Science-fiction writing and popular movies, from *2001: A Space Odyssey* (1968) to *Avengers: Age of Ultron* (2015), have speculated about artificial intelligence (AI) that exceeds the expectations of its creators and escapes their control, eventually outcompeting and enslaving humans or targeting them for extinction.
3. Conflict between humans and AI is front and center in AMC's sci-fi series *Humans*, which returned for its third season on Tuesday (June 5). In the new episodes, conscious synthetic humans face hostile people who treat them with suspicion, fear and hatred. Violence roils as Synths find themselves fighting for not only basic rights but their very survival, against those who view them as less than human and as a dangerous threat.
4. Even in the real world, not everyone is ready to welcome AI with open arms. In recent years, as computer scientists have pushed the boundaries of what AI can accomplish, leading figures in technology and science have warned about the looming dangers that AI may pose to humanity, even suggesting that AI capabilities could doom the human race.
5. But why are people so unnerved by the idea of AI?

An "Existential Threat"

6. Elon Musk is one of the prominent voices that has raised red flags about AI. In July 2017, Musk told attendees at a meeting of the National Governors Association, "I have exposure to

the very cutting-edge AI, and I think people should be really concerned about it."

7. "I keep sounding the alarm bell," Musk added. "But until people see robots going down the street killing people, they don't know how to react, because it seems so ethereal."

8. Earlier, in 2014, Musk had labeled AI "our biggest existential threat" .

9. Physicist Stephen Hawking, who died March 14, also expressed concerns about malevolent AI, telling the BBC in 2014 that "the development of full AI could spell the end of the human race".

10. It's also less than reassuring that some programmers—particularly those with MIT Media Lab in Cambridge, Massachusetts—seem determined to prove that AI can be terrifying.

11. A neural network called "Nightmare Machine", introduced by MIT computer scientists in 2016, transformed ordinary photos into ghoulish, unsettling hellscapes. An AI that the MIT group dubbed "Shelley" composed scary stories, trained on 140,000 tales of horror that Reddit users posted in the forum r/nosleep.

12. "We are interested in how AI induces emotions—fear, in this particular case," Manuel Cebrian, a research manager at MIT Media Lab, previously told Live Science in an e-mail about Shelley's scary stories.

Fear and Loathing

13. Negative feelings about AI can generally be divided into two categories: the idea that AI will become conscious and seek to destroy us, and the notion that immoral people will use AI for evil purposes, Kilian Weinberger, an associate professor in the Department of Computer Science at Cornell University, told Live Science.

14. "One thing that people are afraid of, is that if super-intelligent AI—more intelligent than us—becomes conscious, it could treat us like lower beings, like we treat monkeys," he said. "That would certainly be undesirable."

15. However, fears that AI will develop awareness and overthrow humanity are grounded in misconceptions of what AI is, Weinberger noted. AI operates under very specific limitations defined by the algorithms that dictate its behavior. Some types of problems map well to AI's skill sets, making certain tasks relatively easy for AI to complete. "But most things do not map to that, and they're not applicable," he said.

16. This means that, while AI might be capable of impressive feats within carefully delineated boundaries—playing a master-level chess game or rapidly identifying objects in images, for example—that's where its abilities end.

17. “AI reaching consciousness—there has been absolutely no progress in research in that area,” Weinberger said. “I don’t think that’s anywhere in our near future.”

18. The other worrisome idea—that an unscrupulous human would harness AI for harmful reasons—is, unfortunately, far more likely, Weinberger added. Pretty much any type of machine or tool can be used for either good or bad purposes, depending on the user’s intent, and the prospect of weapons harnessing AI is certainly frightening and would benefit from strict government regulation, Weinberger said.

19. Perhaps, if people could put aside their fears of hostile AI, they would be more open to recognizing its benefits, Weinberger suggested. Enhanced image-recognition algorithms, for example, could help dermatologists identify moles that are potentially cancerous, while self-driving cars could one day reduce the number of deaths from auto accidents, many of which are caused by human error, he told Live Science.

20. But in the *Humans* world of self-aware Synths, fears of conscious AI spark violent confrontations between Synths and people, and the struggle between humans and AI will likely continue to unspool and escalate—during the current season, at least.

(865 words)

Glossary:

algorithm *n.* a precise rule (or set of rules) specifying how to solve some problem 算法

delineate *v.* determine the essential quality of 划定边界

dermatologist *n.* a doctor who specializes in the physiology and pathology of the skin 皮肤科医生

escalate *v.* increase in extent or intensity 升级

ethereal *adj.* as impalpable or intangible as air 缥缈的

ghoulish *adj.* suggesting the horror of death and decay 恐怖的

malevolent *adj.* arising from intense ill will or hatred 有恶意的

speculate *v.* to believe especially on uncertain or tentative grounds 推测

unscrupulous *adj.* an ethical or moral principle that inhibits action 肆无忌惮的

Text B

Why Computers Will Never Be Truly Conscious

1. Many advanced AI projects say they are working toward building a conscious machine, based

on the idea that brain functions merely encode and process multisensory information. The assumption goes, then, that once brain functions are properly understood, it should be possible to program them into a computer. Microsoft recently announced that it would spend $1 billion on a project to do just that.

2. So far, though, attempts to build supercomputer brains have not even come close. A multi-billion-dollar European project that began in 2013 is now largely understood to have failed. That effort has shifted to look more like a similar but less ambitious project in the U.S., developing new software tools for researchers to study brain data, rather than simulating a brain.

3. Some researchers continue to insist that simulating neuroscience with computers is the way to go. Others, like me, view these efforts as doomed to failure because we do not believe consciousness is computable. Our basic argument is that brains integrate and compress multiple components of an experience, including sight and smell—which simply can't be handled in the way today's computers sense, process and store data.

Brains Don't Operate Like Computers

4. Living organisms store experiences in their brains by adapting neural connections in an active process between the subject and the environment. By contrast, a computer records data in short-term and long-term memory blocks. That difference means the brain's information handling must also be different from how computers work.

5. The mind actively explores the environment to find elements that guide the performance of one action or another. Perception is not directly related to the sensory data: A person can identify a table from many different angles, without having to consciously interpret the data and then ask its memory if that pattern could be created by alternate views of an item identified some time earlier.

6. Another perspective on this is that the most mundane memory tasks are associated with multiple areas of the brain—some of which are quite large. Skill learning and expertise involve reorganization and physical changes, such as changing the strengths of connections between neurons. Those transformations cannot be replicated fully in a computer with a fixed architecture.

Computation and Awareness

7. In my own recent work, I've highlighted some additional reasons that consciousness is not

computable.

8. A conscious person is aware of what they're thinking, and has the ability to stop thinking about one thing and start thinking about another—no matter where they were in the initial train of thought. But that's impossible for a computer to do. More than 80 years ago, pioneering British computer scientist Alan Turing showed that there was no way ever to prove that any particular computer program could stop on its own—and yet that ability is central to consciousness.

9. His argument is based on a trick of logic in which he creates an inherent contradiction: Imagine there were a general process that could determine whether any program it analyzed would stop. The output of that process would be either "yes, it will stop" or "no, it won't stop". That's pretty straightforward. But then Turing imagined that a crafty engineer wrote a program that included the stop-checking process, with one crucial element: an instruction to keep the program running if the stop-checker's answer was "yes, it will stop".

10. Running the stop-checking process on this new program would necessarily make the stop-checker wrong: If it determined that the program would stop, the program's instructions would tell it not to stop. On the other hand, if the stop-checker determined that the program would not stop, the program's instructions would halt everything immediately. That makes no sense—and the nonsense gave Turing his conclusion, that there can be no way to analyze a program and be entirely absolutely certain that it can stop. So it's impossible to be certain that any computer can emulate a system that can definitely stop its train of thought and change to another line of thinking—yet certainty about that capability is an inherent part of being conscious.

11. Even before Turing's work, German quantum physicist Werner Heisenberg showed that there was a distinct difference in the nature of the physical event and an observer's conscious knowledge of it. This was interpreted by Austrian physicist Erwin Schrödinger to mean that consciousness cannot come from a physical process, like a computer's, that reduces all operations to basic logic arguments.

12. These ideas are confirmed by medical research findings that there are no unique structures in the brain that exclusively handle consciousness. Rather, functional MRI imaging shows that different cognitive tasks happen in different areas of the brain. This has led neuroscientist Semir Zeki to conclude that "consciousness is not a unity, and that there are instead many consciousnesses that are distributed in time and space". That type of limitless brain capacity isn't the sort of challenge a finite computer can ever handle.

(779 words)

Glossary:

emulate *v.* imitate the function of (another system), as by modifying the hardware or the software （计算机）仿真

halt *v.* cause to stop 停止

inherent *adj.* existing as an essential constituent or characteristic 内在的

integrate *v.* make into a whole or make part of a whole 整合

mundane *adj.* very ordinary and therefore not interesting 平凡的

perception *n.* becoming aware of something via the senses 认识能力

quantum *n.* the smallest discrete quantity of some physical property that a system can possess 量子

simulate *v.* create a representation or model of 仿真

transformation *n.* a qualitative change 转型

Exercises

Ⅰ. Mindmap and Summary

Read the passages and finish the mindmap and summary exercises.

Text A

Analyze the supporting ideas of each part of the passage, and complete the following mindmap.

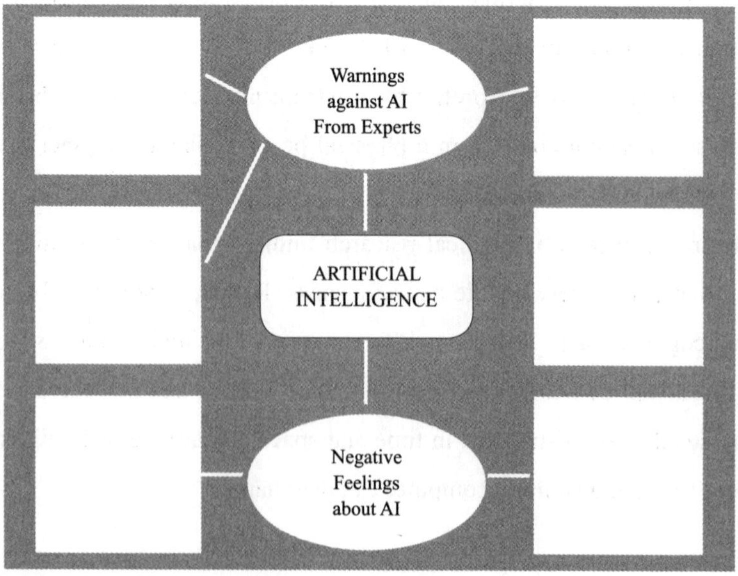

Text B

Analyze the information given in the passage, and write a summary.

____

____

____

____

____

Ⅱ. Paraphrasing

Interpret the following sentences in your own words.

1. But buried in the joke is a seed of unease.
2. Physicist Stephen Hawking, who died March 14, also expressed concerns about malevolent AI, telling the BBC in 2014 that "the development of full AI could spell the end of the human race".
3. Perhaps, if people could put aside their fears of hostile AI, they would be more open to recognizing its benefits, Weinberger suggested.
4. Another perspective on this is that the most mundane memory tasks are associated with multiple areas of the brain—some of which are quite large.
5. These ideas are confirmed by medical research findings that there are no unique structures in the brain that exclusively handle consciousness.

Ⅲ. Brief Answering

Answer the following questions based on information given in the two passages.

1. According to Text A, what does "existential threat" mean?
2. What are the negative feelings about AI?
3. What is Weinberger's attitude toward AI?
4. According to Text B, how do the brain and the computer differ in storing information?
5. How does medical research assist in proving that the computer does possess consciousness?

Ⅳ. Evaluation and Critical Thinking

Based on your analysis of the two passages, discuss in groups and present your ideas on the following items.

1. What function does paragraph 2 serve in the development of Text A?
2. How are the two negative views about AI refuted in Text A?
3. What is the function of mentioning the Synths at the end of Text A?
4. Explain how logic is used in the argument concerning whether the computer has consciousness in Text B?
5. Based on what you have learned in Text A and Text B, what kind of attitude should we take on AI?

Ⅴ. True/False Checking

Based on your understanding of the passages, decide whether each statement is true or false.

Text A

1. Conflict between humans and AI is a central issue in science and technology today.
2. People hold diverse views on AI.
3. AI is the biggest existential threat to mankind.
4. Misunderstanding of the essence of AI gives rise to the fear of it.
5. Only when people are unbiased about AI, could they come to the recognition of its contributions.

Text B

6. Once brain functions are properly understood, it should be possible to program them into a computer.
7. The mechanism of the brain handling information is different from that of the computer.
8. Spontaneous cease of the train of thought is central to consciousness.
9. Alan Turing proved that the computer had no consciousness on the base of a paradox.
10. Erwin Schrödinger explained the difference between the nature of the physical event and an observer's conscious knowledge of it before Turing.

Ⅵ. Translation

Translate the following sentences into Chinese.

1. In recent years, as computer scientists have pushed the boundaries of what AI can accomplish, leading figures in technology and science have warned about the looming dangers that AI may

pose to humanity, even suggesting that AI capabilities could doom the human race. (Para. 4, Text A)

2. "I keep sounding the alarm bell," Musk added. "But until people see robots going down the street killing people, they don't know how to react, because it seems so ethereal." (Para. 7, Text A)

3. Pretty much any type of machine or tool can be used for either good or bad purposes, depending on the user's intent, and the prospect of weapons harnessing AI is certainly frightening and would benefit from strict government regulation, Weinberger said. (Para. 18, Text A)

4. The assumption goes, then, that once brain functions are properly understood, it should be possible to program them into a computer. (Para. 1, Text B)

5. So it's impossible to be certain that any computer can emulate a system that can definitely stop its train of thought and change to another line of thinking—yet certainty about that capability is an inherent part of being conscious. (Para. 10, Text B)

Additional Reading

Artificial Intelligence: Friendly or Frightening?

1. It's a Saturday morning in June at the Royal Society in London. Computer scientists, public figures and reporters have gathered to witness or take part in a decades-old challenge. Some of the participants are flesh and blood; others are silicon and binary. Thirty human judges sit down at computer terminals, and begin chatting. The goal? To determine whether they're talking to a computer program or a real person.

2. The event, organized by the University of Reading, was a rendition of the so-called Turing test, developed 65 years ago by British mathematician and cryptographer Alan Turing as a way to assess whether a machine is capable of intelligent behavior indistinguishable from that of a human. The recently released film *The Imitation Game*, about Turing's efforts to crack the German Enigma code during World War II, is a reference to the scientist's own name for his test.

3. In the London competition, one computerized conversation program, or chatbot, with the personality of a 13-year-old Ukrainian boy named Eugene Goostman, rose above and beyond

the other contestants. It fooled 33 percent of the judges into thinking it was a human being. At the time, contest organizers and the media hailed the performance as an historic achievement, saying the chatbot was the first machine to "pass" the Turing test.

4. When people think of artificial intelligence (AI)—the study of the design of intelligent systems and machines—talking computers like Eugene Goostman often come to mind. But most AI researchers are focused less on producing clever conversationalists and more on developing intelligent systems that make people's lives easier—from software that can recognize objects and animals, to digital assistants that cater to, and even anticipate, their owners' needs and desires.

5. But several prominent thinkers, including the famed physicist Stephen Hawking and billionaire entrepreneur Elon Musk, warn that the development of AI should be cause for concern.

Thinking Machines

6. The notion of intelligent automata, as friend or foe, dates back to ancient times.

7. "The idea of intelligence existing in some form that's not human seems to have a deep hold in the human psyche," said Don Perlis, a computer scientist who studies AI at the University of Maryland, College Park.

8. Reports of people worshipping mythological human likenesses and building humanoid automatons date back to the days of ancient Greece and Egypt, Perlis told Live Science. AI has also featured prominently in pop culture, from the sentient computer HAL 9000 in Stanley Kubrick's *2001: A Space Odyssey* to Arnold Schwarzenegger's robot character in *The Terminator* films.

9. Since the field of AI was officially founded in the mid-1950s, people have been predicting the rise of conscious machines, Perlis said. Inventor and futurist Ray Kurzweil, recently hired to be a director of engineering at Google, refers to a point in time known as "the singularity", when machine intelligence exceeds human intelligence. Based on the exponential growth of technology according to Moore's Law (which states that computing processing power doubles approximately every two years), Kurzweil has predicted the singularity will occur by 2045.

10. But cycles of hype and disappointment—the so-called "winters of AI"—have characterized the history of AI, as grandiose predictions failed to come to fruition. The University of Reading Turing test is just the latest example: Many scientists dismissed the Eugene

Goostman performance as a parlor trick; they said the chatbot had gamed the system by assuming the persona of a teenager who spoke English as a foreign language. (In fact, many researchers now believe it's time to develop an updated Turing test.)

11. Nevertheless, a number of prominent science and technology experts have expressed worry that humanity is not doing enough to prepare for the rise of artificial general intelligence, if and when it does occur. Earlier this week, Hawking issued a dire warning about the threat of AI.

12. "The development of full artificial intelligence could spell the end of the human race," Hawking told the BBC, in response to a question about his new voice recognition system, which uses AI to predict intended words. (Hawking has a form of the neurological disease amyotrophic lateral sclerosis, ALS or Lou Gehrig's disease, and communicates using specialized speech software.)

13. And Hawking isn't alone. Musk told an audience at MIT that AI is humanity's "biggest existential threat". He also once tweeted, "We need to be super careful with AI. Potentially more dangerous than nukes."

14. In March, Musk, Facebook CEO Mark Zuckerberg and actor Ashton Kutcher jointly invested $40 million in the company Vicarious FPC, which aims to create a working artificial brain. At the time, Musk told CNBC that he'd like to "keep an eye on what's going on with AI", adding, "I think there's potentially a dangerous outcome there."

15. But despite the fears of high-profile technology leaders, the rise of conscious machines—known as "strong AI" or "general AI"—is likely a long way off, many researchers argue.

16. "I don't see any reason to think that as machines become more intelligent—which is not going to happen tomorrow—they would want to destroy us or do harm," said Charlie Ortiz, head of AI at the Burlington, Massachusetts-based software company Nuance Communications. "Lots of work needs to be done before computers are anywhere near that level," he said.

Machines with Benefits

17. AI is a broad and active area of research, but it's no longer the sole province of academics; increasingly, companies are incorporating AI into their products.

18. And there's one name that keeps cropping up in the field: Google. From smartphone assistants to driverless cars, the Bay Area-based tech giant is gearing up to be a major player in the future of AI.

19. Google has been a pioneer in the use of machine learning—computer systems that can learn from data, as opposed to blindly following instructions. In particular, the company uses a set of machine-learning algorithms, collectively referred to as "deep learning", that allow a computer to do things such as recognize patterns from massive amounts of data.

20. For example, in June 2012, Google created a neural network of 16,000 computers that trained itself to recognize a cat by looking at millions of cat images from YouTube videos, *The New York Times* reported. (After all, what could be more uniquely human than watching cat videos?)

21. The project, called Google Brain, was led by Andrew Ng, an AI researcher at Stanford University who is now the chief scientist for the Chinese search engine Baidu, which is sometimes referred to as "China's Google".

22. Today, deep learning is a part of many products at Google and at Baidu, including speech recognition, Web search and advertising, Ng told Live Science in an email.

23. Current computers can already complete many tasks typically performed by humans. But possessing humanlike intelligence remains a long way off, Ng said. "I think we're still very far from the singularity. This isn't a subject that most AI researchers are working toward."

24. Gary Marcus, a cognitive psychologist at NYU who has written extensively about AI, agreed. "I don't think we're anywhere near human intelligence [for machines] ," Marcus told Live Science. In terms of simulating human thinking, "we are still in the piecemeal era."

25. Instead, companies like Google focus on making technology more helpful and intuitive. And nowhere is this more evident than in the smartphone market.

Artifiial Intelligence in Your Pocket

26. In the 2013 movie *Her*, actor Joaquin Phoenix's character falls in love with his smartphone operating system, "Samantha," a computer-based personal assistant who becomes sentient. The film is obviously a product of Hollywood, but experts say that the movie gets at least one thing right: Technology will take on increasingly personal roles in people's daily lives, and will learn human habits and predict people's needs.

27. Anyone with an iPhone is probably familiar with Apple's digital assistant Siri, first introduced as a feature on the iPhone 4S in October 2011. Siri can answer simple questions, conduct Web searches and perform other basic functions. Microsoft's equivalent is Cortana, a digital assistant available on Windows phones. And Google has the Google app, available for

Android phones or iPhones, which bills itself as providing "the information you want, when you need it".

28. For example, Google Now can show traffic information during your daily commute, or give you shopping list reminders while you're at the store. You can ask the app questions, such as "should I wear a sweater tomorrow?" and it will give you the weather forecast. And, perhaps a bit creepily, you can ask it to "show me all my photos of dogs" (or "cats" "sunsets" or even person's name), and the app will find photos that fit that description, even if you haven't labeled them as such.

29. Given how much personal data from users Google stores in the form of emails, search histories and cloud storage, the company's deep investments in AI may seem disconcerting. For example, AI could make it easier for the company to deliver targeted advertising, which some users already find unpalatable. And AI-based image recognition software could make it harder for users to maintain anonymity online.

30. But the company, whose motto is "Don't be evil", claims it can address potential concerns about its work in AI by conducting research in the open and collaborating with other institutions, company spokesman Jason Freidenfelds told Live Science. In terms of privacy concerns, specifically, he said, "Google goes above and beyond to make sure your information is safe and secure," calling data security a "top priority".

31. While a phone that can learn your commute, answer your questions or recognize what a dog looks like may seem sophisticated, it still pales in comparison with a human being. In some areas, AI is no more advanced than a toddler. Yet, when asked, many AI researchers admit that the day when machines rival human intelligence will ultimately come. The question is, are people ready for it?

Taking Artificial Intelligence Seriously

32. In the 2014 film *Transcendence*, actor Johnny Depp's character uploads his mind into a computer, but his hunger for power soon threatens the autonomy of his fellow humans.

33. Hollywood isn't known for its scientific accuracy, but the film's themes don't fall on deaf ears. In April, when *Trancendence* was released, Hawking and fellow physicist Frank Wilczek, cosmologist Max Tegmark and computer scientist Stuart Russell published an op-ed in *The Huffington Post* warning of the dangers of AI.

34. "It's tempting to dismiss the notion of highly intelligent machines as mere science fiction,"

Hawking and others wrote in the article. "But this would be a mistake, and potentially our worst mistake ever."

35. Undoubtedly, AI could have many benefits, such as helping to aid the eradication of war, disease and poverty, the scientists wrote. Creating intelligent machines would be one of the biggest achievements in human history, they wrote, but it "might also be [the] last". Considering that the singularity may be the best or worst thing to happen to humanity, not enough research is being devoted to understanding its impacts, they said.

36. As the scientists wrote, "Whereas the short-term impact of AI depends on who controls it, the long-term impact depends on whether it can be controlled at all."

Unit Two Neurotechnology

Critical Reading

Text A

The Future of Mind Control

1. Electrodes implanted in the brain help alleviate symptoms like the intrusive tremors associated with Parkinson's disease. But current probes face limitations due to their size and inflexibility. "The brain is squishy and these implants are rigid," said Shaun Patel. About four years ago, when he discovered Charles M. Lieber's ultra-flexible alternatives, he saw the future of brain-machine interfaces.
2. In a recent perspective titled *Precision Electronic Medicine*, published in *Nature Biotechnology*, Patel, a faculty member at the Harvard Medical School and Massachusetts General Hospital, and Lieber, the Joshua and Beth Friedman University Professor, argue that neurotechnology is on the cusp of a major renaissance. Throughout history, scientists have blurred discipline lines to tackle problems larger than their individual fields. The Human Genome Project, for example, convened international teams of scientists to map human genes faster than otherwise possible.
3. "The next frontier is really the merging of human cognition with machines," Patel said. He and Lieber see mesh electronics as the foundation for those machines, a way to design personalized electronic treatment for just about anything related to the brain.
4. "Everything manifests in the brain fundamentally. Everything. All your thoughts, your perceptions, any type of disease," Patel said.
5. Scientists can pinpoint the general areas of the brain where decision-making, learning, and emotions originate, but tracing behaviors to specific neurons is still a challenge. Right now, when the brain's complex circuitry starts to misbehave or degrade due to psychiatric illnesses like addiction or Obsessive-Compulsive Disorder, neurodegenerative diseases like Parkinson's

or Alzheimer's, or even natural aging, patients have only two options for medical intervention: drugs or, when thosc fail, implanted electrodes.

6. Drugs like L-dopa can quiet the tremors that prevent someone with Parkinson's from performing simple tasks like dressing and eating. But because drugs affect more than just their target, even common L-dopa side effects can be severe, ranging from nausea to depression to abnormal heart rhythms.

7. When drugs no longer work, FDA-approved electrodes can provide relief through Deep Brain Stimulation. Like a pace maker, a battery pack set beneath the clavicle sends automated electrical pulses to two brain implants. Lieber said each electrode "looks like a pencil. It's big".

8. During implantation, Parkinson's patients are awake, so surgeons can calibrate the electrical pulses. Dial the electricity up, and the tremors calm. "Almost instantly, you can see the person regain control of their limbs," Patel said. "It blows my mind."

9. But, like with L-dopa, the large electrodes stimulate more than their intended targets, causing sometimes severe side effects like speech impediments. And, over time, the brain's immune system treats the stiff implants as foreign objects: Neural immune cells (glia cells) engulf the perceived invader, displacing or even killing neurons and reducing the device's ability to maintain treatment.

10. In contrast, Lieber's mesh electronics provoke almost no immune response. With close, long-term proximity to the same neurons, the implants can collect robust data on how individual neurons communicate over time or, in the case of neurological disorders, fail to communicate. Eventually, such technology could track how specific neural subtypes talk, too, all of which could lead to a cleaner, more precise map of the brain's communication network.

11. With higher resolution targets, future electrodes can act with greater precision, eliminating unwanted side effects. If that happens, Patel said, they could be tuned to treat any neurological disorder. And, unlike current electrodes, Lieber's have already demonstrated a valuable trick of their own: They encourage neural migration, potentially guiding newborn neurons to damaged areas, like pockets created by stroke.

12. "The potential for it is outstanding," Patel said. "In my own mind, I see this at the level of what started with the transistor or telecommunications."

13. The potential reaches beyond therapeutics: Adaptive electrodes could provide heightened control over prosthetic or even paralyzed limbs. In time, they could act like neural substitutes, replacing damaged circuitry to re-establish broken communication networks and recalibrate based on live feedback. "If you could actually interact in a precise and long-term way and also provide feedback information," Lieber said, "you could really communicate with the

brain in the same way that the brain is communicating within itself."

14. A few major technology companies are also eager to champion brain-machine interfaces. Some, like Elon Musk's Neuralink, which plans to give paralyzed patients the power to work computers with their minds, are focused on assistive applications. Others have broader plans: Facebook wants people to text by imaging the words, and Brian Johnson's Kernel hopes to enhance cognitive abilities.

15. During his postdoctoral studies, Patel saw how just a short pulse of electricity—no more than 500 milliseconds of stimulation—could control a person's ability to make a safe or impulsive decision. After a little zap, subjects who almost always chose the risky bet, instead went with the safe option. "You would have no idea that it's happened," Patel said. "You're unaware of it. It's beyond your conscious awareness."

16. Such power demands intense ethical scrutiny. For people struggling to combat addiction or obsessive-compulsive disorder, an external pulse regulator could significantly improve their quality of life. But, companies that operate those regulators could access their client's most personal data—their thoughts. And, if enhanced learning and memory are for sale, who gets to buy a better brain? "One does need to be a little careful about the ethics involved if you're trying to make a superhuman," Lieber said. "Being able to help people is much more important to me at this time."

17. Mesh electronics still have several major challenges to overcome: scaling up the number of implanted electrodes, processing the data flood those implants deliver, and feeding that information back into the system to enable live recalibration.

18. "I always joke in talks that I'm doing this because my memory has gotten a little worse than it used to be," Lieber said. "That's natural aging. But does it have to be that way? What if you could correct it?" If he and Patel succeed in galvanizing researchers around mesh electronics, the question might not be if but when.

(1,002 words)

Glossary:

alleviate *v.* provide physical relief 缓解

convene *v.* call together 召集

engulf *v.* cover completely 吞噬

galvanize *v.* stimulate to action 刺激

impediment *n.* something immaterial that interferes with or delays action or progress 障碍

merging *n.* joining together as one 融合

prosthetic *adj.* of or related a replacement for a part of the body 义肢的

proximity *n.* the property of being close together 接近

scrutiny *n.* examining something closely 详细审查

therapeutics *n.* branch of medicine concerned with the treatment of disease 治疗学

Text B

First-Ever Successful Mind-Controlled Robotic Arm without Brain Implants

1. A team of researchers from Carnegie Mellon University, in collaboration with the University of Minnesota, has made a breakthrough in the field of noninvasive robotic device control. Using a noninvasive brain-computer interface (BCI), researchers have developed the first-ever successful mind-controlled robotic arm exhibiting the ability to continuously track and follow a computer cursor.
2. Being able to noninvasively control robotic devices using only thoughts will have broad applications, in particular benefiting the lives of paralyzed patients and those with movement disorders.
3. BCIs have been shown to achieve good performance for controlling robotic devices using only the signals sensed from brain implants. When robotic devices can be controlled with high precision, they can be used to complete a variety of daily tasks. Until now, however, BCIs successful in controlling robotic arms have used invasive brain implants. These implants require a substantial amount of medical and surgical expertise to correctly install and operate, not to mention cost and potential risks to subjects, and as such, their use has been limited to just a few clinical cases.
4. A grand challenge in BCI research is to develop less invasive or even totally noninvasive technology that would allow paralyzed patients to control their environment or robotic limbs using their own "thoughts". Such noninvasive BCI technology, if successful, would bring such much needed technology to numerous patients and even potentially to the general population.
5. However, BCIs that use noninvasive external sensing, rather than brain implants, receive "dirtier" signals, leading to current lower resolution and less precise control. Thus, when using only the brain to control a robotic arm, a noninvasive BCI doesn't stand up to using implanted devices. Despite this, BCI researchers have forged ahead, their eye on the prize of a less-or non-invasive technology that could help patients everywhere on a daily basis.

6. Bin He, Trustee Professor and Department Head of Biomedical Engineering at Carnegie Mellon University, is achieving that goal, one key discovery at a time.
7. "There have been major advances in mind controlled robotic devices using brain implants. It's excellent science," says He. "But noninvasive is the ultimate goal. Advances in neural decoding and the practical utility of noninvasive robotic arm control will have major implications on the eventual development of noninvasive neurorobotics."
8. Using novel sensing and machine learning techniques, He and his lab have been able to access signals deep within the brain, achieving a high resolution of control over a robotic arm. With noninvasive neuroimaging and a novel continuous pursuit paradigm, He is overcoming the noisy EEG signals leading to significantly improve EEG-based neural decoding, and facilitating real-time continuous 2D robotic device control.
9. Using a noninvasive BCI to control a robotic arm that's tracking a cursor on a computer screen, for the first time ever, He has shown in human subjects that a robotic arm can now follow the cursor continuously. Whereas robotic arms controlled by humans noninvasively had previously followed a moving cursor in jerky, discrete motions—as though the robotic arm was trying to "catch up" to the brain's commands—now, the arm follows the cursor in a smooth, continuous path.
10. In a paper published in *Science Robotics*, the team established a new framework that addresses and improves upon the "brain" and "computer" components of BCI by increasing user engagement and training, as well as spatial resolution of noninvasive neural data through EEG source imaging.
11. The paper, *Noninvasive neuroimaging enhances continuous neural tracking for robotic device control*, shows that the team's unique approach to solving this problem not only enhanced BCI learning by nearly 60% for traditional center-out tasks, it also enhanced continuous tracking of a computer cursor by over 500%.
12. The technology also has applications that could help a variety of people, by offering safe, noninvasive "mind control" of devices that can allow people to interact with and control their environments. The technology has, to date, been tested in 68 able-bodied human subjects (up to 10 sessions for each subject), including virtual device control and controlling of a robotic arm for continuous pursuit. The technology is directly applicable to patients, and the team plans to conduct clinical trials in the near future.
13. "Despite technical challenges using noninvasive signals, we are fully committed to bringing this safe and economic technology to people who can benefit from it," says He. "This work represents an important step in noninvasive brain-computer interfaces, a technology which

someday may become a pervasive assistive technology aiding everyone, like smartphones."

14. This work was supported in part by the National Center for Complementary and Integrative Health, National Institute of Neurological Disorders and Stroke, National Institute of Biomedical Imaging and Bioengineering, and National Institute of Mental Health.

(779 words)

Glossary:

cursor *n.* indicator consisting of a movable spot of light 光标
decode *v.* convert code into ordinary language 解码
forge *v.* move ahead steadily 前进
facilitate *v.* make easier 促进
paradigm *n.* a standard or typical example 范例
pervasive *adj.* spreading throughout 遍布的
resolution *n.* the number of pixels per square inch on a computer-generated display 分辨率
subject *n.* a person who is subjected to experimental or other observational procedures 被试
substantial *adj.* fairly large 大量的

Exercises

Ⅰ. Summary and Mindmap

Read the passages and finish the summary and mindmap exercises.

Text A

Analyze the information given in the passage, and write a summary.

____

____

____

____

____

Text B

Analyze the supporting ideas of each part of the passage, and complete the following mindmap.

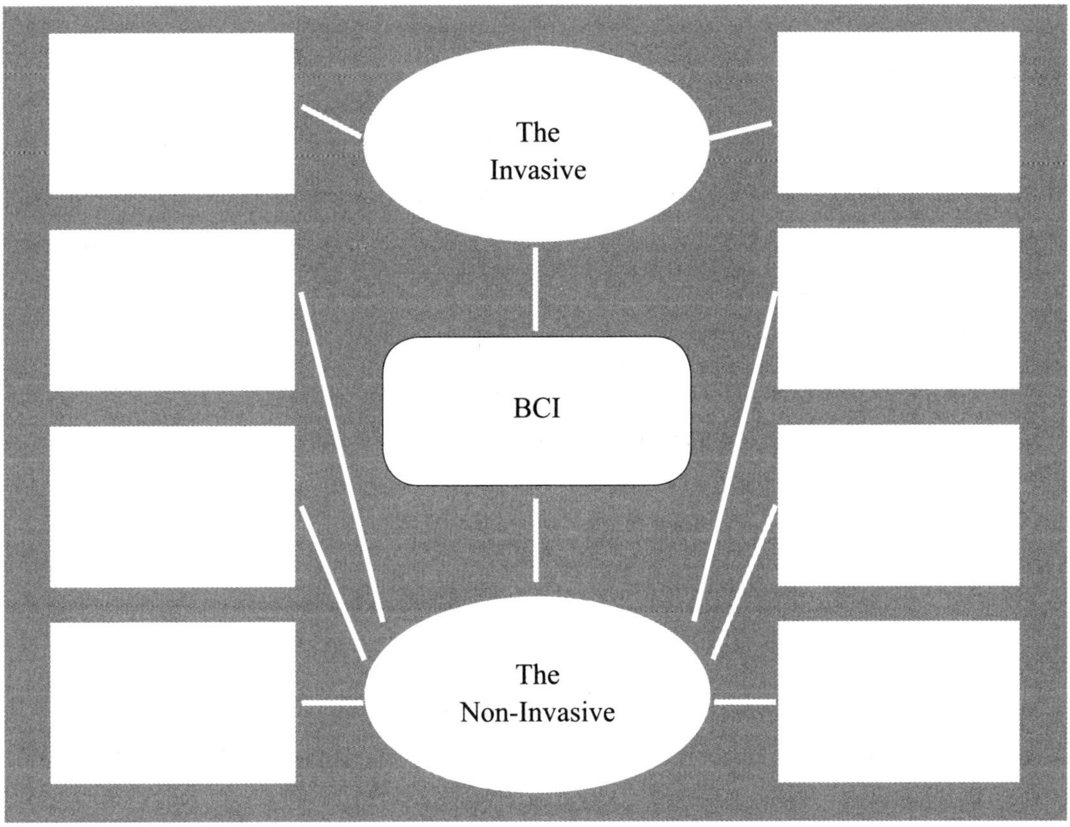

Ⅱ. Paraphrasing

Interpret the following sentences in your own words.

1. Neurotechnology is on the cusp of a major renaissance. Throughout history, scientists have blurred discipline lines to tackle problems larger than their individual fields.
2. Scientists can pinpoint the general areas of the brain where decision-making, learning, and emotions originate, but tracing behaviors to specific neurons is still a challenge.
3. But, companies that operate those regulators could access their client's most personal data—their thoughts. And, if enhanced learning and memory are for sale, who gets to buy a better brain?
4. These implants require a substantial amount of medical and surgical expertise to correctly install and operate, not to mention cost and potential risks to subjects, and as such, their use

has been limited to just a few clinical cases.

5.Thus, when using only the brain to control a robotic arm, a noninvasive BCI doesn't stand up to using implanted devices. Despite this, BCI researchers have forged ahead, their eye on the prize of a less-or non-invasive technology that could help patients everywhere on a daily basis.

Ⅲ. Brief Answering

Answer the following questions based on information given in the two passages.

1. According to Patel, what is the next stage of development in mind control?
2. What are the problems with drugs and implanted electrodes as methods to treat neurodegenerative diseases?
3. What are the technical hurdles to overcome for mesh electronics?
4. What are the problems with BCIs that use noninvasive external sensing?
5. What is important for the eventual development of noninvasive neurorobotics?

Ⅳ. Evaluation and Critical Thinking

Based on your analysis of the two passages, discuss in groups and present your ideas on the following items.

1. What is the function of paragraph 2 in the development of Text A?
2. Apart from what the few major technique companies plan for as presented in Text A, what do you think could be other possibilities of brain-machine interface?
3. How would you comment on the writing of the last paragraph of Text A?
4. What is the function of mentioning all the organizations supporting the work at the end of Text B?
5. Based on what you have learned in Text A and Text B, what kind of ethical issues might ensue with the development in BCI?

Ⅴ. True/False Checking

Based on your understanding of the passages, decide whether each statement is true or false.

Text A

1. Cooperation between scientists specializing in different fields is not rare.
2. Scientists cannot pinpoint the specific neurons determining behaviors.

3. Large implanted electrodes may be under the attack from the immune system.
4. The development in this field of technology is not free from ethical issues.
5. Mesh electronics are far from perfect.

Text B

6. The cost of invasive brain implants is very high.
7. Machine learning techniques assist in accessing signals deep within the brain.
8. Robotic arms controlled by humans noninvasively cannot follow moving cursors smoothly.
9. More involvement of users is conducive to improvement of the components of BCI.
10. The technology of BCI may someday enable people to interact with their environment.

Ⅵ. Translation

Translate the following sentences into Chinese.

1. Electrodes implanted in the brain help alleviate symptoms like the intrusive tremors associated with Parkinson's disease. But current probes face limitations due to their size and inflexibility. (Para. 1, Text A)
2. But because drugs affect more than just their target, even common L-dopa side effects can be severe, ranging from nausea to depression to abnormal heart rhythms. (Para. 6, Text A)
3. Such power demands intense ethical scrutiny. For people struggling to combat addiction or obsessive-compulsive disorder, an external pulse regulator could significantly improve their quality of life. (Para. 16, Text A)
4. A grand challenge in BCI research is to develop less invasive or even totally noninvasive technology that would allow paralyzed patients to control their environment or robotic limbs using their own "thoughts". (Para. 4, Text B)
5. Despite this, BCI researchers have forged ahead, their eye on the prize of a less-or non-invasive technology that could help patients everywhere on a daily basis. (Para. 5, Text B)

Additional Reading

AI Can Now Decode Words Directly from Brain Waves

1. Neuroscientists are teaching computers to read words straight out of people's brains.

2. Kelly Servick, writing for *Science*, reported this week on three papers posted to the preprint server bioRxiv in which three different teams of researchers demonstrated that they could decode speech from recordings of neurons firing. In each study, electrodes placed directly on the brain recorded neural activity while brain-surgery patients listened to speech or read words out loud. Then, researchers tried to figure out what the patients were hearing or saying. In each case, researchers were able to convert the brain's electrical activity into at least somewhat-intelligible sound files.
3. The first paper, posted to bioRxiv on Oct. 10, 2018, describes an experiment in which researchers played recordings of speech to patients with epilepsy who were in the middle of brain surgery. (The neural recordings taken in the experiment had to be very detailed to be interpreted. And that level of detail is available only during the rare circumstances when a brain is exposed to the air and electrodes are placed on it directly, such as in brain surgery.)
4. As the patients listened to the sound files, the researchers recorded neurons firing in the parts of the patients' brains that process sound. The scientists tried a number of different methods for turning that neuronal firing data into speech and found that "deep learning"—in which a computer tries to solve a problem more or less unsupervised—worked best. When they played the results through a vocoder, which synthesizes human voices, for a group of 11 listeners, those individuals were able to correctly interpret the words 75 percent of the time.
5. You can listen to audio from this experiment here.
6. The second paper, posted Nov. 27, 2018, relied on neural recordings from people undergoing surgery to remove brain tumors. As the patients read single-syllable words out loud, the researchers recorded both the sounds coming out of the participants' mouths and the neurons firing in the speech-producing regions of their brains. Instead of training computers deeply on each patient, these researchers taught an artificial neural network to convert the neural recordings into audio, showing that the results were at least reasonably intelligible and similar to the recordings made by the microphones. (The audio from this experiment is here but has to be downloaded as a zip file.)
7. The third paper, posted Aug. 9, 2018, relied on recording the part of the brain that converts specific words that a person decides to speak into muscle movements. While no recording from this experiment is available online, the researchers reported that they were able to reconstruct entire sentences (also recorded during brain surgery on patients with epilepsy) and that people who listened to the sentences were able to correctly interpret them on a multiple choice test (out of 10 choices) 83 percent of the time. That experiment's method relied on identifying the patterns involved in producing individual syllables, rather than whole words.

8. The goal in all of these experiments is to one day make it possible for people who've lost the ability to speak (due to amyotrophic lateral sclerosis or similar conditions) through a computer-to-brain interface. However, the science for that application isn't there yet.
9. Interpreting the neural patterns of a person just imagining speech is more complicated than interpreting the patterns of someone listening to or producing speech, Science reported. (However, the authors of the second paper said that interpreting the brain activity of someone imagining speech may be possible.)
10. It's also important to keep in mind that these are small studies. The first paper relied on data taken from just five patients, while the second looked at six patients and the third only three. And none of the neural recordings lasted more than an hour.
11. Still, the science is moving forward, and artificial-speech devices hooked up directly to the brain seem like a real possibility at some point down the road.

Unit Three Environmental Science

Critical Reading

Text A

10 Studies That Revealed the Great Global Amphibian Die-off—and Some Possible Solutions

1. In our planet's sixth great mass extinction event, amphibians are among the hardest hit.
2. Amphibian species have been facing a steep decline for decades, in large part because of a fungus, climate change, and environment disruption. As many as one-third of the world's 6,300 amphibian species are threatened with extinction, and researchers fear their loss could wreak havoc on our ecosystem and food webs. Here are landmark studies that have defined the problems and—we hope—will help humans to figure out how to save their froggy friends.

Ⅰ. The Long-Term Perspective

3. In August 2008 two researchers at University of California at Berkeley published a meta-analysis titled *Are we in the midst of the sixth mass extinction.* The global assessment highlighted the threat of chytridiomycosis, an infectious, rapidly spreading disease caused by a waterborne fungus.

Ⅱ. The Seminal Study

4. The study *Status and Trends of Amphibian Declines and Extinctions Worldwide* [subscription required] showed one-third of the world's 6,300 amphibian species are threatened with extinction (compared with just 12 percent of all bird species and 23 percent of mammal species). The authors concluded that scientists must begin captive breeding.

Ⅲ. Problems with Captive Breeding

5. A September 2008 study published in *Current Biology* said captive breeding program accidentally introduced the chytrid fungus that causes chytridiomycosis into Mallorca in 1991; an endangered frog species was housed in the same room as a group of toads, and the frogs spread the chytrid fungus to the toads. The fungus was not known at this time, so health screening of the toads did not reveal the problem.

Ⅳ. Extinction Rates

6. According to a 2007 study in the *Journal of Herpetology*, amphibian species are becoming extinct 211 times faster than the "normal extinction rate", the standard rate of extinction in history before humans became a primary contributing factor. And if you count those species "in imminent danger of extinction", that rate climbs to a whopping 45,474 times faster than normal.

Ⅴ. The Global-Warming Theory

7. It is unclear how the chytrid fungus spreads. A 2006 study in *Nature* [subscription required] blamed global warming, for creating ideal conditions shifts temperatures to those more agreeable to the fungus' growth and reproduction—between 63°F and 77°F. But there were many outspoken skeptics of this claim (see study #6).

Ⅵ. Skeptics of the Global-Warming Theory

8. A March 2008 PLoS Biology paper said there is "no evidence to support" the global-warming hypothesis. Instead, researchers said, the pattern of the fungus' spread was typical of an emerging infectious disease; they call their theory the "spreading pathogen hypothesis". The authors suggested governments and environmental agencies can help prevent the fungus' spread by regulating potential infection routes, such as the ornamental plant and aquarium wildlife trade.

Ⅶ. Problems with Insecticides

9. October 2008's *Ecological Applications* [subscription required] study suggests malathion, the most common insecticide in the U.S., can devastate tadpole populations even at doses too small to kill individual tadpoles. Researchers created simulated ponds in 300-gallon outdoor

tanks, placed tadpoles inside, and exposed the ponds to no malathion, a single moderately concentrated dose, or low concentrations in weekly doses similar to the exposure tadpoles experience in (human-altered) nature. Even the small amounts of malathion set off a chain of events that caused a decline of tadpoles' primary food source: bottom-dwelling algae. Consequently, half the tadpoles in the experiment did not reach maturity and would have died in nature.

Ⅷ. A Fungus-Free Country

10. Oddly, none of the many known amphibian species in Madagascar have been driven to extinction; the island also shows no signs of the chytrid fungus, according to a May 2008 PLoS Biology paper. Because the amphibians in Madagascar are doing so well, the authors argue the region is one of the best places to focus future research efforts. They hope to find out what helps amphibians in Madagascar thrive, as they suffer steep declines elsewhere.

Ⅸ. Beating the Fungus

11. Introducing probiotic bacteria into the ecosystem could help lessen the effects of chytridiomycosis, according to research presented in June 2008 at the 108th General Meeting of the American Society for Microbiology. The tests indicated that adding pedobacter, a bacterial species that occurs naturally on the skin of red-backed salamanders, to the skin of mountain yellow-legged frogs decreased chytrid's deadly effects.

Ⅹ. Evolving More Slowly than the Environment

12. A May 2007 study in *BioScience* attributed amphibians' decline to their inability to adapt to the current rapid rate of global change. The authors noted the aforementioned pesticide pollution and chytrid infections, as well as habitat loss and UV-B light exposure that causes mutations in amphibian eggs. Amphibians are particularly vulnerable because they have permeable skin, ability to live on both land and water, and eggs without shells. Perhaps most detrimental is their complex life cycle, which makes evolution an even slower process.

(805 words)

Glossary:

squarium *n.* a tank or pool or bowl filled with water for keeping live fish and underwater animals 水族馆

concentration *n.* bring or becoming gathered 集中

havoc *n.* violent and needless disturbance 破坏

imminent *adj.* close in time; about to occur 逼近的

mutation *n.* any event that changes genetic structure; any alteration in the inherited nucleic acid sequence of the genotype of an organism 突变

pathogen *n.* any disease-producing agent (especially a virus or bacterium or other microorganism) 病原体

screening *n.* testing objects or persons in order to identify those with particular characteristics 筛查

simulated *adj.* reproduced or made to resemble 仿真的

Text B

Wildlife Conservation 2.0

1. Nothing pushes a species to extinction like wiping out its habitat. Consider the Hawaiian Islands: They were originally covered in trees, but by the 1950s three-quarters of the islands' natural forests had been destroyed to make way for animal pastures and crops. Many other habitats were overrun by introduced pigs and rats. The effect on Hawaii's indigenous species was devastating: In the last 200 years, 28 species of birds alone were wiped out, including the large Kauai thrush. Once widespread throughout the Hawaiian Islands, this thrush has not been seen since 1989. It is considered extinct by the World Conservation Union.
2. Conservation biologists face an increasingly difficult job of preserving habitats and, with them, global biodiversity. But Hugh Possingham, an ecologist and mathematician at the University of Queensland in Australia, has developed revolutionary software that will make their work easier and more effective.
3. Traditionally, biologists have drawn up priority lists of places that should be preserved. Sounds straightforward—except that different biologists favor different lists, each list driven by different criteria. One might rank a location according to the overall number of threatened species there, while another ranks locations based on the number of species that are unique to that area. Which list should an organization follow? The most popular list to have emerged,

first proposed in the late 1980s by conservation biologist Norman Myers, pinpoints "biodiversity hot spots"—those places with the greatest number of unique species facing the most severe threats, such as the tropical Andes and the Horn of Africa.

4. A new software-based approach may be the key to saving thousands of species.

5. Possingham questions the conventional wisdom that severely threatened places deserve the most attention, and he sees a better path to preservation. "A consequence of our approach is that you do not spend the most money on the most endangered species or the most endangered regions," he says. "You balance cost and biodiversity and threats."

6. For example, last September Possingham, Kerrie Wilson (a biologist at the University of Queensland), and a team of researchers assessed the cost and outcomes of various conservation actions in 39 "Mediterranean" ecoregions identified by the World Wildlife Foundation (WWF). These regions—in places like California, South Africa, Chile, and Australia—are among the world's most threatened. Analysis showed that to save the most biodiversity for the buck, scientists might do best to spend money on relatively cheap interventions (such as weed control in a Chilean forest, where weed removal means native trees do not have to compete for nutrients in the soil) and eschew more expensive investments in areas such as Australia's Jarrah-Karri forest, even though it has the highest vertebrate diversity of all the Mediterranean regions analyzed and is home to rare marsupials. The goal is to save more species on the whole, even if they are less newsworthy or photogenic.

7. Possingham began developing this unconventional way of thinking in 1994, when he was on sabbatical at Imperial College London and watched biologists scrambling to try to figure out what to save. He was amazed to see that when they drew up their priority lists, they neglected a crucial factor: cost. Well-grounded in math, Possingham began constructing models that performed cost-efficiency analyses of different conservation schemes, ultimately encoding his work into Marxan, a software program written by a Ph.D. student named Ian Ball and first released in 1999. Since then, Possingham has continued to incorporate new factors into his models, including information about the types of threats that species face, the cost of interventions to combat these threats, and the ability to account for how threats and interventions change over time.

8. In practice, Marxan is a tool into which conservationists and policymakers can enter information about their local environment—the distribution of flora and fauna, for example, or the economic value of a patch of land. Based on these data, Marxan designs nature reserves that cost as little as possible to create and maintain, while meeting whatever conservation criteria the user has established; this might mean creating the smallest possible nature reserve

that still represents every type of plant life in a given region.

9. Many conservation organizations and governments around the world have enthusiastically adopted Marxan to design and manage protected areas. The Australian government, for example, recently used Possingham's analysis to guide a series of major conservation decisions. Marxan helped identify regions off Australia's northeast coast that collectively maximized biological diversity in the Great Barrier Reef Marine Park, leading to the rezoning of the park boundaries. The government also used Marxan in designating 50 million hectares of new reserves in other parts of the country.

10. Not everyone raves about Possingham's work. Some claim his software-driven approach is at times unnecessary. Conservation ecologist Stuart Pimm of Duke University thinks that Possingham's models make sense in places like Australia, where there is still a lot of intact biodiversity; he has reservations about its use in places where biodiversity is fast declining. For instance, Pimm and a small group of other scientists are now buying up cattle pastures in Brazil to try to connect fragments of highly diverse—and highly threatened—coastal forests. Pimm calls this action so "obvious" that it requires no methodical cost-benefit analysis. "When you've got a lot of land to play with," Pimm says, "it makes sense to think of these formalized processes [like Possingham's], but in practice, in areas that are particularly badly degraded, you don't have a lot of choices."

11. To many others, though, Marxan's process is part of the appeal. "For years people have sat around with maps and pens and drawn lines on the maps and said, 'We should protect this and protect that'", says Ray Nias, conservation director for the WWF-Australia, based in Sydney. "What Hugh has done is to make that a mathematical and logical process rather than an intuitive one. It's far more sophisticated and robust than the old way of doing things."

12. Possingham and his colleagues are currently working on making Marxan faster and easier to use and adding additional routines to consider the effects of catastrophes like hurricanes. Not a bad thing, if we are to save as many as possible of the 16,306 species currently listed as threatened by the World Conservation Union.

(1,040 words)

Glossary:

catastrophe *n.* an event resulting in great loss and misfortune 大灾难

conservationist *n.* someone who works to protect the environment from destruction or pollution 自然资源保护论者

ecoregion *n*. an area defined by its environmental conditions, esp. climate, landforms, and soil characteristics 生态区域

eschew *v*. avoid and stay away from deliberately 避开

hectare *n*. (abbreviated ‘ha’) a unit of surface area equal to 10,000 square meters 公顷

intervention *n*. the act of getting involved (as to mediate a dispute) 干预

marsupial *n*. mammals of which the females have a pouch (the marsupium) containing the teats where the young are fed and carried 有袋类动物

methodical *adj*. characterized by method and orderliness 有条不紊的

robust *adj*. strong enough to withstand or overcome intellectual challenges or adversity 强大的

Exercises

Ⅰ. Mindmap and Outline

Read the passages and finish the mindmap and outline exercises.

Text A

Analyze and classify the studies presented in the passage, figure out the interrelationship between them, and then finish the following mindmap. Part of the information has been given.

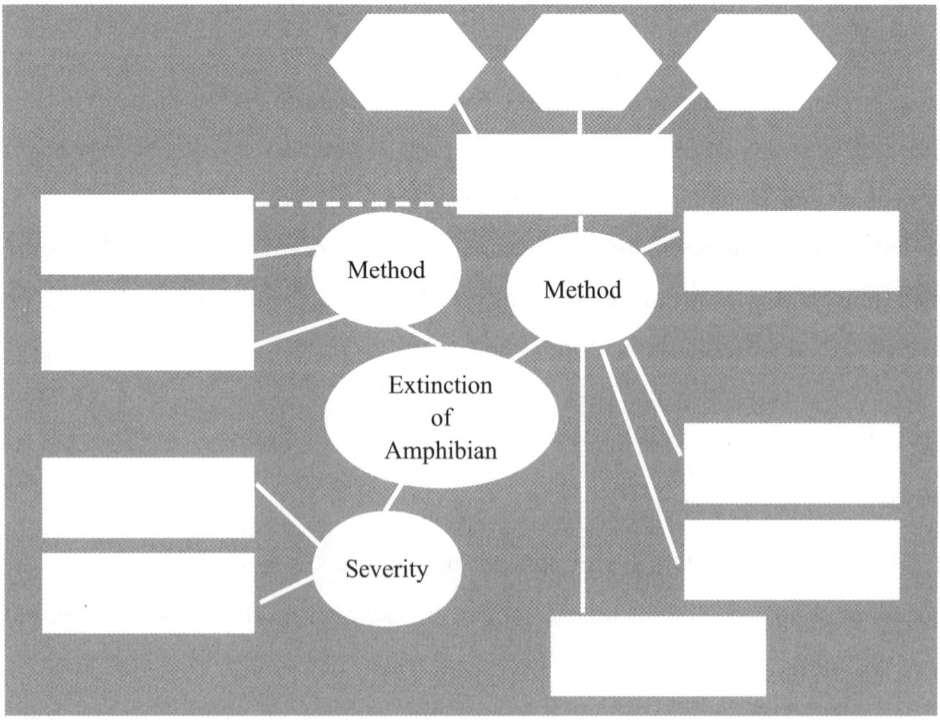

Text B

Analyze the main idea of each part and find out how the writer organizes this passage. Figure out the logical connection between each part, and work out how coherence and cohesion are achieved.

Para. 1 ~ 2	(1)________________________________
Para. 3	(2)__________ way of producing priority lists and its problems
Para. 4 ~ 8	A new software has been worked out to deal with the issue
Para. 9 ~ 11	(3) __________ toward the software
Para. 12	(4) __________ of the software

Ⅱ. Paraphrasing

Interpret the following sentences in your own words.

1. As many as one-third of the world's 6,300 amphibian species are threatened with extinction, and researchers fear their loss could wreak havoc on our ecosystem and food webs.
2. The authors suggested governments and environmental agencies can help prevent the fungus' spread by regulating potential infection routes, such as the ornamental plant and aquarium wildlife trade.
3. Even the small amounts of malathion set off a chain of events that caused a decline of tadpoles' primary food source: bottom-dwelling algae. Consequently, half the tadpoles in the experiment did not reach maturity and would have died in nature.
4. Possingham questions the conventional wisdom that severely threatened places deserve the most attention, and he sees a better path to preservation.
5. In practice, Marxan is a tool into which conservationists and policymakers can enter information about their local environment—the distribution of flora and fauna, for example, or the economic value of a patch of land.

Ⅲ. Brief Answering

Answer the following questions based on information given in the two passages.

1. According to Text A, what is the problem with captive breeding of amphibians?
2. Why does the application of insecticide give rise to extinction of amphibians?
3. What is the purpose of introducing probiotic bacteria into the ecosystem?

4. What are the causes for the extinction of amphibians presented in Text A?
5. How is Possingham's method different from the way biologists traditionally draw up priority lists of places that should be preserved?
6. How do people differ in attitudes towards the usage of Marxan?

Ⅳ. Evaluation and Critical Thinking

Based on your analysis of the two passages, discuss in groups and present your ideas on the following items.

1. Based on information presented in Text A and Text B, what are the factors leading to extinction of species? How to tackle the problem?
2. According to Text A, what aspects concerning the chytrid fungus have been probed into?
3. Based on the analysis of the ten studies and their interrelationship presented in Text A, if you were an ecologist, what suggestions would you offer for future studies in this area?
4. Based on the analysis of the ten studies and their interrelationship presented in Text A, if you were a biologist, what suggestions would you offer for future studies in this area?
5. Why do people have diverse attitudes toward the use of Marxan? And what is the function of presenting the response from Tay Nias?
6. What do learn from the writing of Text B?

Ⅴ. True/False Checking

Based on your understanding of the passages, decide whether each statement is true or false.

Text A

1. Extinction of amphibians could cause damage to ecosystem.
2. We are now in the midst of the sixth mass extinction.
3. Extinction of amphibians is caused by transmission of the chytrid fungus from frogs to toads.
4. Decline in bottom-dwelling algae may lead to the extinction of toads.
5. Certain bacteria are conducive to combating the disease caused by the chytrid fungus.

Text B

6. Eliminating habitats of species poses the greatest threat to species.
7. The most feasible criterion to draw up priority list is to pinpoint the places with the greatest

number of unique species facing the most sever threats.

8. Possingham constructed models and wrote the software program.
9. Marxan serves well in designing nature reserves.
10. Compared with conventional ways, Marxan is more scientific and effective.

Ⅵ. Translation

Translate the following sentences into Chinese.

1. As many as one-third of the world's 6,300 amphibian species are threatened with extinction, and researchers fear their loss could wreak havoc on our ecosystem and food webs. (Para. 2, Text A)
2. The authors suggested governments and environmental agencies can help prevent the fungus' spread by regulating potential infection routes, such as the ornamental plant and aquarium wildlife trade. (Para. 8, Text A)
3. Amphibians are particularly vulnerable because they have permeable skin, ability to live on both land and water, and eggs without shells. Perhaps most detrimental is their complex life cycle, which makes evolution an even slower process. (Para. 12, Text A)
4. Analysis showed that to save the most biodiversity for the buck, scientists might do best to spend money on relatively cheap interventions (such as weed control in a Chilean forest, where weed removal means native trees do not have to compete for nutrients in the soil) and eschew more expensive investments in areas such as Australia's Jarrah-Karri forest, even though it has the highest vertebrate diversity of all the Mediterranean regions analyzed and is home to rare marsupials. (Para. 6, Text B)
5. Not everyone raves about Possingham's work. Some claim his software-driven approach is at times unnecessary. Conservation ecologist Stuart Pimm of Duke University thinks that Possingham's models make sense in places like Australia, where there is still a lot of intact biodiversity; he has reservations about its use in places where biodiversity is fast declining. (Para. 10, Text B)

Additional Reading

The Reality of Climate Change: 10 Myths Busted (Abridged)

Dynamic Earth

1. Earth is a dynamic sphere and, it turns out, so is the planet's climate, otherwise known as the long-term trend of global weather conditions. It's no wonder questions and myths abound about what exactly is going on in the atmosphere, in the oceans and on land. How can we tell our orb is actually warming and whether humans are to blame? Here's a look at what scientists know and don't know about some seemingly murky statements on Earth's climate.

Climate Has Changed Before

2. Myth: Even before SUVs and other greenhouse-gas spewing technologies, Earth's climate was changing, so humans can't be responsible for today's global warming.
3. Science: Climate changes in the past suggest that our climate reacts to energy input and output, such that if the planet accumulates more heat than it gives off global temperatures will rise. It's the driver of this heat imbalance that differs.
4. Currently, CO_2 is imposing an energy imbalance due to the enhanced greenhouse effect. Past climate change actually provides evidence for our climate's sensitivity to CO_2.

But It's Cold Outside!

5. Myth: The planet can't be warming when my front yard is covered in several feet of snow. This winter has been one of the chilliest, how is that possible in a warming world?
6. Science: Local temperatures taken as individual data points have nothing to do with the long-term trend of global warming. These local ups and downs in weather and temperature can hide a slower-moving uptick in long-term climate. To get a real bead on global warming, scientists rely on changes in weather over a long period of time. To find climate trends you need to look at how weather is changing over a longer time span. Looking at high and low temperature data from recent decades shows that new record highs occur nearly twice as often as new record lows.

Climate Is Cooling

7. Myth: Global warming has stopped and the Earth has begun to cool.

8. Science: The last decade, 2000—2009, was the hottest on record, according to Skeptical Science. Big blizzards and abnormally chilly weather often raise the question: How can global warming be occurring when it's snowing outside? Global warming is compatible with chilled weather. "For climate change, it is the long-term trends that are important; measured over decades or more, and those long term trends show that the globe is still, unfortunately, warming," according to Skeptical Science.

The Sun Is to Blame

9. Myth: Over the past few hundred years, the sun's activity, including the number of sunspots, has increased, causing the world to get warmer.

10. Science: In the last 35 years of global warming, the sun has shown a slight cooling trend, while the climate has been heating up, scientists say. In the past century, solar activity can explain some of the increase in global temperatures, but a relatively small amount. (Solar activity refers to the activity of the sun's magnetic field and includes magnetic field-powered sunspots and solar flares.)

Not Everyone Agrees

11. Myth: There's no consensus on whether the planet is actually warming.

12. Science: About 97 percent of climate scientists agree that human-made global warming is happening. "In the scientific field of climate studies—which is informed by many different disciplines—the consensus is demonstrated by the number of scientists who have stopped arguing about what is causing climate change—and that's nearly all of them," according to Skeptical Science, a website dedicated to explaining the science of global warming.

Carbon Dioxide (CO_2) Is Not a Pollutant

13. Myth: Rick Santorum, GOP presidential nominee, summed up this argument in the news when he said: "The dangers of carbon dioxide? Tell that to a plant, how dangerous carbon dioxide is," he told the Associated Press.

14. Science: While it is true that plants photosynthesize, and therefore take up carbon dioxide as a way of forming energy with the help of the sun and water, this gas is both a direct pollutant

(think acidification of oceans) and more importantly is linked to the greenhouse effect. When heat energy gets released from Earth's surface, some of that radiation is trapped by greenhouse gases like CO_2; the effect is what makes our planet comfy temperature-wise, but too much and you get global warming.

Climate Scientists Are Conspiring to Push "Global Warming"

15. Myth: Thousands of e-mails between climate scientists leaked in November 2009 (dubbed Climategate) revealed a cover-up of data that conflicted with research showing the Earth is warming.
16. Science: Yes, a hacker did access and release emails and documents from the University of East Anglia server. But there was no cover-up; a number of investigations were launched, including two independent reviews set up by the university: the Independent Climate Change E-mails Review (ICCER) and the independent Scientific Appraisal Panel (SAP). The investigations cleared the researchers involved with the e-mails of scientific misconduct, and found no evidence of a cover-up.

Don't Worry, It's Not That Bad

17. Myth: Some have pointed to human history as evidence that warm periods are good for people, while the cold, unstable stints have been catastrophic.
18. Science: Climate scientists say any positives are far outweighed by the negative impacts of global warming on agriculture, human health, the economy and the environment. For instance, according to one 2007 study, a warming planet may mean an increased growing season in Greenland; but it also means water shortages, more frequent and more intense wildfires and expanding deserts.

Antarctica Is Gaining Ice

19. Myth: Ice covering much of Antarctica is expanding, contrary to the belief that the ice cap is melting due to global warming.
20. Science: The argument that ice is expanding on Antarctica omit the fact that there's a difference between land ice and sea ice, climate scientists say. "If you are talking about the Antarctic ice sheet, we expect some gain in accumulation in the interior due to warmer, more moisture-laden air, but increased calving/ice loss at the periphery, primarily due to warming southern oceans," climate scientist Michael Mann, of Pennsylvania State University, told

Live Sciencc. Thc nct change in ice mass is the difference between this accumulation and peripheral loss. “Models traditionally have projected that this difference doesn’t become negative (i.e. net loss of Antarctic ice sheet mass) for several decades,” Mann said, adding that detailed gravimetric measurements, which looks at changes in Earth’s gravity over spots to estimate, among other things, ice mass. These measurements, Mann said, suggest the Antarctic ice sheet is already losing mass and contributing to sea level rise.

Climate Models Are Unreliable

21. Myth: Models are full of “fudge factors” or assumptions that make them fit with data collected in today’s climate; there’s no way to know if those same assumption can be made in a world with increased carbon dioxide.
22. Science: Models have successfully reproduced global temperatures since 1900, by land, in the air and the oceans. “Models are simply a formalization of our best understanding of the processes that govern the atmosphere, the oceans, the ice sheets, etc.,” Mann said. He added that certain processes, such as how clouds will respond to changes in the atmosphere and the warming or cooling effect of clouds, are uncertain and different modeling groups make different assumptions about how to represent these processes.

Unit Four Cybersecurity

Critical Reading

Text A

Hackers Could Kill More People than a Nuclear Weapon

1. People around the world may be worried about nuclear tensions rising, but I think they're missing the fact that a major cyberattack could be just as damaging—and hackers are already laying the groundwork.
2. With the U.S. and Russia pulling out of a key nuclear weapons pact—and beginning to develop new nuclear weapons—plus Iran tensions and DPRK again test-launching missiles, the global threat to civilization is high. Some fear a new nuclear arms race.
3. That threat is serious—but another could be as serious, and is less visible to the public. So far, most of the well-known hacking incidents, even those with foreign government backing, have done little more than steal data. Unfortunately, there are signs that hackers have placed malicious software inside U.S. power and water systems, where it's lying in wait, ready to be triggered. The U.S. military has also reportedly penetrated the computers that control Russian electrical systems.

Many Intrusions Already

4. As someone who studies cybersecurity and information warfare, I'm concerned that a cyberattack with widespread impact, an intrusion in one area that spreads to others or a combination of lots of smaller attacks, could cause significant damage, including mass injury and death rivaling the death toll of a nuclear weapon.
5. Unlike a nuclear weapon, which would vaporize people within 100 feet and kill almost everyone within a half-mile, the death toll from most cyberattacks would be slower. People might die from a lack of food, power or gas for heat or from car crashes resulting from

a corrupted traffic light system. This could happen over a wide area, resulting in mass injury and even deaths.

6. This might sound alarmist, but look at what has been happening in recent years, in the U.S. and around the world.

7. In early 2016, hackers took control of a U.S. treatment plant for drinking water, and changed the chemical mixture used to purify the water. If changes had been made—and gone unnoticed—this could have led to poisonings, an unusable water supply and a lack of water.

8. In 2016 and 2017, hackers shut down major sections of the power grid in Ukraine. This attack was milder than it could have been, as no equipment was destroyed during it, despite the ability to do so. Officials think it was designed to send a message. In 2018, unknown cybercriminals gained access throughout the United Kingdom's electricity system; in 2019 a similar incursion may have penetrated the U.S. grid.

9. In August 2017, a Saudi Arabian petrochemical plant was hit by hackers who tried to blow up equipment by taking control of the same types of electronics used in industrial facilities of all kinds throughout the world. Just a few months later, hackers shut down monitoring systems for oil and gas pipelines across the U.S. This primarily caused logistical problems—but it showed how an insecure contractor's systems could potentially cause problems for primary ones.

10. The FBI has even warned that hackers are targeting nuclear facilities. A compromised nuclear facility could result in the discharge of radioactive material, chemicals or even possibly a reactor meltdown. A cyberattack could cause an event similar to the incident in Chernobyl. That explosion, caused by inadvertent error, resulted in 50 deaths and evacuation of 120,000 and has left parts of the region uninhabitable for thousands of years into the future.

Mutual Assured Destruction

11. My concern is not intended to downplay the devastating and immediate effects of a nuclear attack. Rather, it's to point out that some of the international protections against nuclear conflicts don't exist for cyberattacks. For instance, the idea of "mutual assured destruction" suggests that no country should launch a nuclear weapon at another nuclear-armed nation: The launch would likely be detected, and the target nation would launch its own weapons in response, destroying both nations.

12. Cyberattackers have fewer inhibitions. For one thing, it's much easier to disguise the source of a digital incursion than it is to hide where a missile blasted off from. Further, cyberwarfare can start small, targeting even a single phone or laptop. Larger attacks might

target businesses, such as banks or hotels, or a government agency. But those aren't enough to escalate a conflict to the nuclear scale.

Nuclear Grade Cyberattacks

13. There are three basic scenarios for how a nuclear grade cyberattack might develop. It could start modestly, with one country's intelligence service stealing, deleting or compromising another nation's military data. Successive rounds of retaliation could expand the scope of the attacks and the severity of the damage to civilian life.
14. In another situation, a nation or a terrorist organization could unleash a massively destructive cyberattack—targeting several electricity utilities, water treatment facilities or industrial plants at once, or in combination with each other to compound the damage.
15. Perhaps the most concerning possibility, though, is that it might happen by mistake. On several occasions, human and mechanical errors very nearly destroyed the world during the Cold War; something analogous could happen in the software and hardware of the digital realm.

Defending Against Disaster

16. Just as there is no way to completely protect against a nuclear attack, there are only ways to make devastating cyberattacks less likely.
17. The first is that governments, businesses and regular people need to secure their systems to prevent outside intruders from finding their way in, and then exploiting their connections and access to dive deeper.
18. Critical systems, like those at public utilities, transportation companies and firms that use hazardous chemicals, need to be much more secure. One analysis found that only about one-fifth of companies that use computers to control industrial machinery in the U.S. even monitor their equipment to detect potential attacks—and that in 40% of the attacks they did catch, the intruder had been accessing the system for more than a year. Another survey found that nearly three-quarters of energy companies had experienced some sort of network intrusion in the previous year.

(990 words)

Glossary:

devastating *adj.* completely destructive 毁灭性的

discharge *v.* give or send out liquid, gas, electric current, etc. 放出

hazardous *adj.* involving risk or danger 危险的

inadvertent *adj.* happening by chance or unexpectedly or unintentionally 疏忽的；非故意的

incursion *n.* the act of entering some territory or domain (often in large numbers) 入侵

intrusion *n.* entrance by force or without permission or welcome 闯入

logistical *adj.* relating to the careful organization of a complicated activity 统筹安排上的

malicious *adj.* intended to harm 恶意的

retaliation *n.* action taken in return for an injury or offense 报复，反击

Text B

Hackers Target 3rd Dimension of Cyberspace: Users' Minds (Adapted)

1. The attacks on the 2016 U.S. presidential election and continuing election-related hacking have happened across all three dimensions of cyberspace—physical, informational and cognitive. The first two are well-known: For years, hackers have exploited hardware and software flaws to gain unauthorized access to computers and networks—and stolen information they've found. The third dimension, however, is a newer target—and a more concerning one.
2. This three-dimensional view of cyberspace comes from Professor Dan Kuehl of the National Defense University, who expressed concern about traditional hacking activities and what they meant for national security. But he also foresaw the potential—now clear to the public at large—that those tools could be used to target people's perceptions and thought processes, too.
3. Some observers suggest that using internet tools for espionage and as fuel for disinformation campaigns is a new form of "hybrid warfare". Their idea is that the lines are blurring between the traditional kinetic warfare of bombs, missiles and guns, and the unconventional, stealthy warfare long practiced against foreigners' "hearts and minds" by intelligence and special forces capabilities.
4. However, I believe this isn't a new form of war at all: Rather, it is the same old strategies taking advantage of the latest available technologies. Just as online marketing companies use sponsored content and search engine manipulation to distribute biased information to the public, governments are using internet-based tools to pursue their agendas. In other words,

they're hacking a different kind of system through social engineering on a grand scale.

Old Goals, New Techniques

5. More than 2,400 years ago, the Chinese military strategist and philosopher Sun Tzu made it an axiom of war that it's best to "subdue the enemy without fighting". Using information—or disinformation, or propaganda—as a weapon can be one way to destabilize a population and disable the target country. In 1984 a former KGB agent who defected to the West discussed this as a long-term process and more or less predicted what's happening in the U.S. now.
6. False social media accounts have been created to simulate political activists, which purported to be associated with the Tennessee Republican Party. Just that one account attracted more than 100,000 followers. The goal was to distribute propaganda, such as captioned photos, posters or short animated graphics, purposely designed to enrage and engage these accounts' followers. Those people would then pass the information along through their own personal social networks.
7. Starting from seeds planted by fakers, including some who claimed to be U.S. citizens, those ideas grew and flourished through amplification by real people. Unfortunately, whether originating from Russia or elsewhere, fake information and conspiracy theories can form the basis for discussion at major partisan media outlets.
8. As ideas with niche online beginnings moved into the traditional mass media landscape, they serve to keep controversies alive by sustaining divisive arguments on both sides. For instance, one Russian troll factory had its online personas host rallies both for and against each of the major candidates in the 2016 presidential election. Though the rallies never took place, the online buzz about them helped inflame divisions in society.
9. The trolls also set up Twitter accounts purportedly representing local news organizations—including defunct ones—to take advantage of Americans' greater trust of local news sources than national ones. These accounts operated for several years—one for the Chicago Daily News, closed since 1978, was created in May 2014 and collected 20,000 followers—passing along legitimate local news stories, likely seeking to win followers' trust ahead of future disinformation campaigns. Shut down before they could fulfill that end, these accounts cleverly aimed to exploit the fact that many Americans' political views cloud their ability to separate fact from opinion in the news.
10. These sorts of activities are functions of traditional espionage: Foment discord and then sit back while the target population becomes distracted arguing among themselves.

Fighting Digital Disinformation Is Hard

11. Analyzing, let alone countering, this type of provocative behavior can be difficult. Russia isn't alone, either: The U.S. tries to influence foreign audiences and global opinions, including through Voice of America online and radio services and intelligence services' activities. And it's not just governments that get involved. Companies, advocacy groups and others also can conduct disinformation campaigns.

12. Unfortunately, laws and regulations are ineffective remedies. Further, social media companies have been fairly slow to respond to this phenomenon. Twitter reportedly suspended more than 70 million fake accounts earlier this summer. That included nearly 50 social media accounts like the fake Chicago Daily News one.

13. Facebook, too, says it is working to reduce the spread of "fake news" on its platform. Yet both companies make their money from users' activity on their sites—so they are conflicted, trying to stifle misleading content while also boosting users' involvement.

Real Defense Happens in the Brain

14. The best protection against threats to the cognitive dimension of cyberspace depends on users' own actions and knowledge. Objectively educated, rational citizens should serve as the foundation of a strong democratic society. But that defense fails if people don't have the skills—or worse, don't use them—to think critically about what they're seeing and examine claims of fact before accepting them as true.

15. American voters expect ongoing Russian interference in U.S. elections. In fact, it appears to have already begun. To help combat that influence, the U.S. Justice Department plans to alert the public when its investigations discover hacking and disinformation relating to the upcoming 2018 midterm elections.

16. These efforts are a good start, but the real solution will begin when people start realizing they're being subjected to this sort of cognitive attack and that it's not all just a hoax.

(946 words)

Glossary:

axiom *n.* a saying that is widely accepted on its own merits 格言

destabilize *v.* make unstable 动摇

discord *n.* the state of not agreeing or sharing opinions 不一致

disinformation *n.* deliberately disseminated misinformation to influence or confuse rivals 虚假信息

espionage *n.* the action of spying 间谍行为

foment *v.* try to stir up public opinion 煽动

hoax *n.* deliberate trickery intended to gain an advantage 恶作剧

kinetic *adj.* relating to the motion of material bodies 运动的

purport *v.* propose or intend 意图

Exercises

Ⅰ. Outline and Mindmap

Read the passages and finish the outline and mindmap exercises.

Text A

Analyze the main idea of each part and find out how the writer organizes this passage. Figure out the logical connection between each part, and work out how coherence and cohesion are achieved.

Para. 1 ~ 3	(1)__________ is just as damaging as nuclear attacks
Para. 4 ~ 10	(2)__________ in many aspects in various countries
Para. 11 ~ 12	Different from nuclear attacks, there's no (3) __________ for cyberattacks
Para. 13 ~ 15	(4) __________ a nuclear grade cyberattack might develop
Para. 16 ~ 18	Methods to (5) __________

Text B

Analyze the supporting ideas of each part of the passage, and present them in the following mindmap. One category has been given.

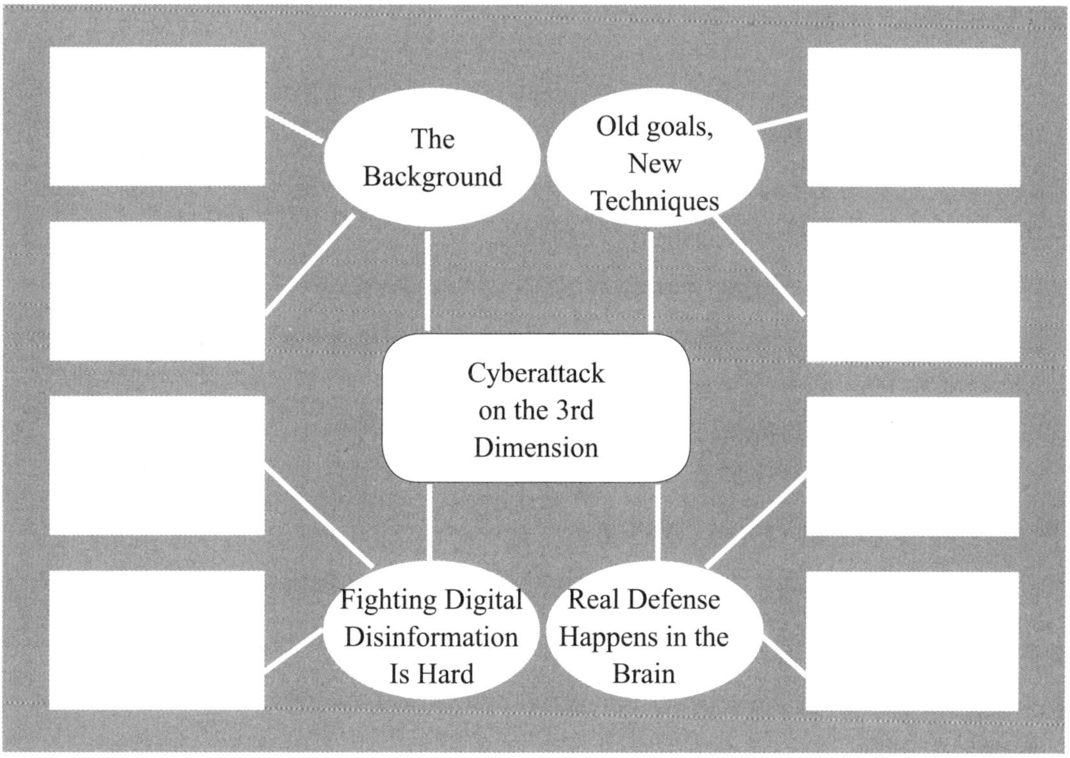

Ⅱ. Paraphrasing

Interpret the following sentences in your own words.

1. Unfortunately, there are signs that hackers have placed malicious software inside U.S. power and water systems, where it's lying in wait, ready to be triggered.
2. Just a few months later, hackers shut down monitoring systems for oil and gas pipelines across the U.S. This primarily caused logistical problems—but it showed how an insecure contractor's systems could potentially cause problems for primary ones.
3. My concern is not intended to downplay the devastating and immediate effects of a nuclear attack. Rather, it's to point out that some of the international protections against nuclear conflicts don't exist for cyberattacks.
4. Starting from seeds planted by fakers, including some who claimed to be U.S. citizens, those ideas grew and flourished through amplification by real people.
5. Russia isn't alone, either: The U.S. tries to influence foreign audiences and global opinions, including through Voice of America online and radio services and intelligence services' activities.

Ⅲ. Brief Answering

Answer the following questions based on information given in the two passages.

1.What areas have been intruded by cyberattackers as presented in Text A?

2.According to Text A, what would happen as a result of the fact that there is no such thing as "mutual assured destruction" in the cyberworld?

3.What are the methods to lessen the effects of devastating cyberattacks?

4.What do you think are the reasons for the fact that cyberattacks on the third dimension deserves more concern, based on the argument in Text B?

5.What does "real defense happens in the brain" mean in Text B?

Ⅳ. Evaluation and Critical Thinking

Based on your analysis of the two passages, discuss in groups and present your ideas on the following items.

1. Based on information presented in Text A and Text B, why are cyberattacks as dangerous as nuclear attacks?
2. What are the purposes for cyberattacks?
3. Distinguish between facts and opinions presented in each text, and then answer the question: Are the authors objective or biased in writing these passages?
4. Can we simply say that it is just or unjust for each side to do what they do in the cases? Why or why not?
5. It is noted in the passages that there are countries executing cyberattacks on each other. Based on the analysis of this fact, answer the questions: What is the implication for us? What is sensible for us to do to safeguard our national cybersecurity?

Ⅴ. True/False Checking

Based on your understanding of the passages, decide whether each statement is true or false.

Text A

1. There are mutual cyberintrusions between the U.S. and Russia.
2. Cyberattacks could result in much higher casualties.
3. Cyberattacks rival nuclear attacks in scale.
4. Nuclear facilities are not the only thing under the threat of cyberattacks.

5. A great majority of companies take no measures to protect their computers from cyberattacks.

Text B

6. The 3rd dimension of cyberspace is related to people's perception.
7. Using internet tools for espionage and as fuel for disinformation campaigns is a new form of warfare.
8. The public's ability to separate fact from opinion in the news plays a role in how cyberattacks work.
9. Objectively educated, rational citizens are the foundation of a strong democratic society.
10. It takes time for the public to understand the whole picture of cyberspace which they are involved in.

Ⅵ. Translation

Translate the following sentences into Chinese.

1. As someone who studies cybersecurity and information warfare, I'm concerned that a cyberattack with widespread impact, an intrusion in one area that spreads to others or a combination of lots of smaller attacks, could cause significant damage, including mass injury and death rivaling the death toll of a nuclear weapon. (Para. 4, Text A)
2. In August 2017, a Saudi Arabian petrochemical plant was hit by hackers who tried to blow up equipment by taking control of the same types of electronics used in industrial facilities of all kinds throughout the world. (Para. 9, Text A)
3. There are three basic scenarios for how a nuclear grade cyberattack might develop. It could start modestly, with one country's intelligence service stealing, deleting or compromising another nation's military data. Successive rounds of retaliation could expand the scope of the attacks and the severity of the damage to civilian life. (Para. 13, Text A)
4. Starting from seeds planted by fakers, including some who claimed to be U.S. citizens, those ideas grew and flourished through amplification by real people. Unfortunately, whether originating from Russia or elsewhere, fake information and conspiracy theories can form the basis for discussion at major partisan media outlets. (Para. 7, Text B)
5. But that defense fails if people don't have the skills—or worse, don't use them—to think critically about what they're seeing and examine claims of fact before accepting them as true. (Para. 14, Text B)

Additional Reading

What Was the World's 1st Cyberattack?

1. Back in November 1988, Robert Tappan Morris, son of the famous cryptographer Robert Morris Sr., was a 20-something graduate student at Cornell who wanted to know how big the internet was—that is, how many devices were connected to it. So he wrote a program that would travel from computer to computer and ask each machine to send a signal back to a control server, which would keep count.
2. The program worked well too well, in fact. Morris had known that if it traveled too fast there might be problems, but the limits he built in weren't enough to keep the program from clogging up large sections of the internet, both copying itself to new machines and sending those pings back. When he realized what was happening, even his messages warning system administrators about the problem couldn't get through.
3. His program became the first of a particular type of cyberattack called "distributed denial of service", in which large numbers of internet-connected devices, including computers, webcams and other smart gadgets, are told to send lots of traffic to one particular address, overloading it with so much activity that either the system shuts down or its network connections are completely blocked.
4. As the chair of the integrated Indiana University Cybersecurity Program, I can report that these kinds of attacks are increasingly frequent today. In many ways, Morris's program, known to history as the "Morris worm", set the stage for the crucial, and potentially devastating, vulnerabilities in what I and others have called the coming "Internet of Everything".

Unpacking the Morris Worm

5. Worms and viruses are similar, but different in one key way: A virus needs an external command, from a user or a hacker, to run its program. A worm, by contrast, hits the ground running all on its own. For example, even if you never open your e-mail program, a worm that gets onto your computer might e-mail a copy of itself to everyone in your address book.
6. In an era when few people were concerned about malicious software and nobody had protective software installed, the Morris worm spread quickly. It took 72 hours for researchers at Purdue and Berkeley to halt the worm. In that time, it infected tens of thousands of systems—about 10 percent of the computers then on the internet. Cleaning up the infection cost hundreds or

thousands of dollars for each affected machine.

7. In the clamor of media attention about this first event of its kind, confusion was rampant. Some reporters even asked whether people could catch the computer infection. Sadly, many journalists as a whole haven't gotten much more knowledgeable on the topic in the intervening decades.

8. Morris wasn't trying to destroy the internet, but the worm's widespread effects resulted in him being prosecuted under the then-new Computer Fraud and Abuse Act. He was sentenced to three years of probation and a roughly US $10,000 fine. In the late 1990s, though, he became a dot-com millionaire—and is now a professor at MIT.

Rising Threats

9. The internet remains subject to much more frequent—and more crippling—DDoS attacks. With more than 20 billion devices of all types, from refrigerators and cars to fitness trackers, connected to the internet, and millions more being connected weekly, the number of security flaws and vulnerabilities is exploding.

10. In October 2016, a DDoS attack using thousands of hijacked webcams—often used for security or baby monitors—shut down access to a number of important internet services along the eastern U.S. seaboard. That event was the culmination of a series of increasingly damaging attacks using a botnet, or a network of compromised devices, which was controlled by software called Mirai. Today's internet is much larger, but not much more secure, than the internet of 1988.

11. Some things have actually gotten worse. Figuring out who is behind particular attacks is not as easy as waiting for that person to get worried and send out apology notes and warnings, as Morris did in 1988. In some cases—the ones big enough to merit full investigations—it's possible to identify the culprits. A trio of college students was ultimately found to have created Mirai to gain advantages when playing the "Minecraft" computer game.

Fighting DDoS Attacks

12. But technological tools are not enough, and neither are laws and regulations about online activity—including the law under which Morris was charged. The dozens of state and federal cybercrime statutes on the books have not yet seemed to reduce the overall number or severity of attacks, in part because of the global nature of the problem.

13. There are some efforts underway in Congress to allow attack victims in some cases to engage

in active defense measures—a notion that comes with a number of downsides, including the risk of escalation—and to require better security for internet-connected devices. But passage is far from assured.

14. There is cause for hope, though. In the wake of the Morris worm, Carnegie Mellon University established the world's first Cyber Emergency Response Team, which has been replicated in the federal government and around the world. Some policymakers are talking about establishing a national cybersecurity safety board, to investigate digital weaknesses and issue recommendations, much as the National Transportation Safety Board does with airplane disasters.

15. More organizations are also taking preventative action, adopting best practices in cybersecurity as they build their systems, rather than waiting for a problem to happen and trying to clean up afterward. If more organizations considered cybersecurity as an important element of corporate social responsibility, they—and their staff, customers and business partners—would be safer.

16. In *3001: The Final Odyssey*, science fiction author Arthur C. Clarke envisioned a future where humanity sealed the worst of its weapons in a vault on the moon—which included room for the most malignant computer viruses ever created. Before the next iteration of the Morris worm or Mirai does untold damage to the modern information society, it is up to everyone—governments, companies and individuals alike—to set up rules and programs that support widespread cybersecurity, without waiting another 30 years.

Unit Five Material Science

Critical Reading

Text A

The Kilogram Isn't What It Used to Be—It's Lighter

1. What I love best about the kilogram is its tangibility, its solid, sculpted form of shiny platinum and iridium. I'm referring to not just any kilogram but the quintessential one that resides here—the actual International Prototype Kilogram, or IPK, created in 1879 as the official standard of mass. It's a smooth cylinder of alloy, only an inch and a half high and an inch and a half in diameter. Though petite, the IPK is necessarily dense; it weighs 2.204,6 pounds. If you went to pick it up, you might think someone had cemented it to the tabletop for a prank. Even if you knew what to expect, its compact heft would still boggle your senses.
2. Of course, they won't let you pick it up. They won't even let you anywhere near it. If you touched it—if you so much as breathed on it—you would change its mass, and then where would we be? That's why the IPK leads such a sheltered life. It is kept under a triple bell jar inside a temperature-and-humidity controlled vault in a secure room within the Parc de Saint-Cloud enclave of the International Bureau of Weights and Measures, or BIPM (Bureau International des Poids et Mesures). Thus protected, it reigns over a world's worth of measurement. Every hill of beans, every lump of coal, every milligram of medication—in short, every quantity of any substance that can be weighed—must be gauged against this object.
3. The IPK is, in and of itself, the International System of Units' definition of mass. Through a complex dissemination protocol, the essence of the kilogram is transferred from the IPK to its counterparts at standards laboratories around the world, and from there to centers of industry and scientific research, ending up in grocery stores, post offices, and bathrooms everywhere.
4. Although I have come to pay my respects to the IPK, I am denied even a glimpse of the

thing. Nor can I see one of its six official copies, for these reside alongside the prototype in guarded seclusion. I must content myself with replicas—with the working standards that fill the ultraclean laboratory of Richard Davis, an American physicist in Paris who for the past 15 years has headed the Mass Section at the BIPM.

5. Gloved for work, Davis wears a lab coat over his street clothes, blue paper bootees over his shoes, and a net over his hair. Around him kilogram weights of various shapes and materials sit on colored plates under glass bell jars, like an assortment of fine cheeses. They have been delivered here from other countries to be reckoned in comparison with the IPK.

6. "That one belongs to Ireland," Davis says, indicating a stainless-steel kilogram on a red dish. Member states—signatories to the Meter Convention—pay dues to the BIPM that cover the cost of periodic checks on their national reference standards.

7. It takes a minimum of four days to calibrate a single kilogram according to the BIPM's cautious regimen of repeated comparison weighings. Visiting kilograms could theoretically go home after a week, but they typically stay in the lab for months, allowing the time it takes them to become thermally stable in their new surroundings, undergo cleaning by the BIPM method, and prove themselves, through repeated trials, to be worthy ambassadors of mass. Given the uncertainty, however minuscule, in every measurement, such repetitions are essential before these national standards can leave with a calibration certificate stating how they compare with the IPK, along with a precise correction factor.

8. En route to or from Paris, the visiting kilograms disdain ordinary transport. Zeina Jabbour, group leader of Mass & Force at the National Institute of Standards and Technology in Gaithersburg, Maryland, recently brought two of the four U.S. kilograms here for calibration. She carried one herself in a specially designed case inside a padded camera bag that was all but handcuffed to her wrist, and she entrusted the other to a colleague who flew on a different plane. ("That way, if something happened to one of us.") Soon after her flight touched down at Charles de Gaulle, she grabbed a taxi straight to the BIPM on the other side of the city for a handover directly to Davis.

9. Before picking up a kilogram with a pair of wide-mouthed forceps called lifters, Davis flicks off suspected specks of dust with a fine-tipped brush. ("My wife paints.") He has modified the artist's brush for his purposes by degreasing its fibers and covering its metal ferrule with plastic, "so if you accidentally hit the kilogram, you won't scratch it." On a balance precise to 10 decimal places, a scratch counts.

10. Davis tests the Irish kilogram in a sealed chamber against three BIPM working standards that are also made of stainless steel. He doesn't weigh it against the platinum-iridium standard,

since stainless weights are only one-third as dense, and therefore three times as large, displacing a much greater quantity of air. "You'd have to make an air buoyancy correction that would amount to almost a tenth of a gram," he explains. "That is huge."

11. Although Davis serves as the IPK's official guardian, even he rarely sees the original prototype, which is too precious and vulnerable to damage to remain in constant use. Over the course of its century-plus lifetime, the IPK has emerged only three times to serve "campaigns" of active duty, most recently in 1988–1992, when it participated in a formal verification of all kilogram prototypes belonging to the 51 Meter Convention member states. On that occasion, however, the IPK itself was found wanting. Despite all the protective protocols and delicate procedures, it had mysteriously changed. No one can say whether the IPK has lost weight (perhaps by the gradual escape of gases trapped inside it from the start) or if most of the prototypes have gained (possibly by accumulating atmospheric contaminants). The difference is approximately 30 micrograms—30 billionths of a kilogram—in a hundred years. (Imagine 30 cents out of a $10 million stack of pennies.)

12. This alarming show of instability is driving global efforts to redefine the kilogram, so that mass need not depend on the safety or stability of some manufactured item stored in a safe. In fact, more than mass hangs in the balance, for the kilogram is tied to three other base units of the International System of Units (SI), namely the ampere, the mole, and the candela. Several more quantities—including density, force, and pressure—are in turn derived from the kilogram.

13. Other 19th-century artifacts of measurement have long since been retired in favor of fundamental constants of nature. In 1983, for example, the platinum-iridium bar that described the length of the meter yielded to a new benchmark: A meter is now defined as the distance light travels in a vacuum in 1/299,792,458 second (a second being the time it takes an atom of cesium-133 to vacillate 9,192,631,770 times between the two hyperfine levels of its ground state). These figures fail to give the average person any real feel for the quantities in question, but to a metrologist—one who specializes in the science of measurement—such equivalences rooted in physics have the advantage of permanence and reproducibility.

14. One invariant vying to replace the IPK is Planck's constant, which could be determined via an experimental device called a watt balance. Alternatively, researchers may successfully express mass in terms of Avogadro's number (which is tied to the unchanging mass of individual atoms), provided they can count the atoms in a crystal of silicon-28.

15. But neither of these complex, costly endeavors is likely to yield a new standard in time for the next meeting of the General Conference of Weights and Measures, scheduled for 2011. For now, the International Prototype Kilogram stands firm on metrology's last frontier.

(1,308 words)

Glossary:

assortment *n.* a collection containing a variety of sorts of things 混合物

benchmark *n.* a standard by which something can be measured or judged 基准点

calibrate *v.* mark (the scale of a measuring instrument) so that it can be read in the desired units 校正刻度

dissemination *n.* the opening of a subject to widespread discussion and debate 传播

hyperfine *adj.* extremely fine or thin, as in a spectral line split into two or more components 超精细的

protocol *n.* forms of ceremony and etiquette observed by diplomats and heads of state 外交礼仪

prototype *n.* a standard or typical example 原型

quintessential *adj.* representing the perfect example of a class or quality 典型的

tangibility *n.* the quality of being perceivable by touch 实体性

vacillate *v.* move or sway in a rising and falling or wavelike pattern 摇摆

Text B

There's a Brand-New Kilogram, And It's Based on Quantum Physics

1. The kilogram isn't a thing anymore. Instead, it's an abstract idea about light and energy.
2. As of today (May 20), physicists have replaced the old kilogram—a 130-year-old, platinum-iridium cylinder weighing 2.2 pounds (1 kilogram) sitting in a room in France—with an abstract, unchanging measurement based on quadrillions of light particles and Planck's constant (a fundamental feature of our universe).
3. In one sense, this is a grand (and surprisingly difficult) achievement. The kilogram is fixed forever now. It can't change over time as the cylinder loses an atom here or an atom there. That means humans could communicate this unit of mass, in terms of raw science, to space aliens. The kilogram is now a simple truth, an idea that can be carried anywhere in the universe without bothering to bring a cylinder with you.

4. And still...so what? Practically speaking, the new kilogram weighs, to within a few parts per billion, exactly as much as the old kilogram did. If you weighed 93 kilograms (204 pounds) yesterday, you'll weigh 93 kilograms today and tomorrow. Only in a few narrow scientific applications will the new definition make any difference.
5. What's really fascinating here isn't that, practically speaking, the way most of us use the kilogram will change. It's how difficult it turned out to be to rigorously define a unit of mass at all.
6. Other fundamental forces have long since been understood in terms of fundamental reality. A second of time? Once, according to the National Institute of Standards and Technology (NIST), it was defined in terms of the swings of a pendulum clock. But now scientists understand a second as the time it takes an atom of cesium-133 to go through 9,192,631,770 cycles of releasing microwave radiation. A meter? That's the distance light travels in 1/299,792,458th of a second.
7. But mass isn't like that. We usually measure kilograms in terms of weight—how much does this thing push down on a scale? But that's a measurement that depends on where you perform the actual weighing. That cylinder in France would weigh much less if you brought it to the moon, and even a tiny bit more or tiny bit less if you brought it to other parts of the Earth.
8. As NIST explains, the new kilogram is based on the fundamental relationship between mass and energy—the relationship partly spelled out in Einstein's $E=mc^2$, which means energy is equal to mass times the speed of light squared. Mass can be converted to energy and vice versa. And, compared with mass, energy is easier to measure and define in discrete terms.
9. That's thanks to another equation, even older than $E=mc^2$. The physicist Max Planck showed in 1900 that $E=hv$, according to NIST. He showed that, on a small enough scale, energy can go up and down, and only in steps. $E=hv$ means that energy is equal to "v"—the frequency of some particle, like a photon—multiplied by "h"—the number $6.626{,}070{,}15\times10^{-34}$ also known as Planck's constant.
10. "v" in $E=hv$ must always be an integer, like 1, 2, 3 or 6,492. No fractions or decimals allowed. So, energy is by its nature discrete, going up and down in steps of "h" ($6.626{,}070{,}15\times10^{-34}$).
11. The new kilogram brings $E=mc^2$ and $E=hv$ together. That enables scientists to define mass in terms of Planck's constant, an unchanging feature of the universe. An international coalition of science labs came together to make the most precise measurements of Planck's constant yet, certain to within just several parts per billion. The new kilogram's mass corresponds to

the energy of 1.475,521,4 times 10^{40} photons that are oscillating at the same frequencies as the cesium-133 atoms used in atomic clocks.

12. It's not the easiest thing to stick on a scale. But, as an idea, it's a lot more portable than a cylinder of platinum-iridium alloy.

(649 words)

Glossary:

coalition *n.* the union of diverse things into one body or form or group 联合，联盟

convert *v.* change 转变

decimal *n.* a number in the decimal system 小数

fraction *n.* the quotient of two rational numbers 分数

oscillate *v.* move or swing from side to side regularly 摆动

pendulum *n.* weight hung on a cord from a fixed point so that it can swing freely 摆，钟摆

quadrillions *n.* the number that is represented as a one followed by 15 zeros 千万亿

release *v.* release (gas or energy) as a result of a chemical reaction or physical decomposition 放出

rigorously *adv.* in a rigidly accurate manner 严格地

Exercises

Ⅰ. Outline and Summary

Read the passages and finish the outline and summary exercises.

Text A

Analyze the main idea of each part and find out how the writer organizes this passage. Figure out the logical connection between each part, and work out how coherence and cohesion are achieved.

Para. 1	The IPK and its (1)_________
Para. 3 ～ 4	How the IPK works as (2)_________
Para. 5 ～ 10	(3) _________ need to be taken in dealing with the IPK
Para. 11	The IPK has (4) _________

Para. 12 ~ 14	(5) _________ of the change in the IPK
Para. 15	The IPK remains (6) _________

Text B

Analyze the passage and figure out the information other than the mere fact that the kilogram is now based on quantum physics. And write a summary of the text.

____

____

____

____

____

Ⅱ. Paraphrasing

Interpret the following sentences in your own words.

1. What I love best about the kilogram is its tangibility, its solid, sculpted form of shiny platinum and iridium.
2. It takes a minimum of four days to calibrate a single kilogram according to the BIPM's cautious regimen of repeated comparison weighings.
3. Over the course of its century-plus lifetime, the IPK has emerged only three times to serve "campaigns" of active duty, most recently in 1988–1992, when it participated in a formal verification of all kilogram prototypes belonging to the 51 Meter Convention member states.
4. The kilogram is now a simple truth, an idea that can be carried anywhere in the universe without bothering to bring a cylinder with you.
5. An international coalition of science labs came together to make the most precise measurements of Planck's constant yet, certain to within just several parts per billion.

Ⅲ. Brief Answering

Answer the following questions based on information given in the two passages.

1. Why are there six official copies of the IPK?
2. What is the implication concerning the brush from Davis' wife, who paints, in Text A?
3. According to Text A, what might be the reasons for the change of the weight of the IPK?
4. How does quantum physics serve in determining the kilogram?

5. What does the last paragraph of Text B imply?

Ⅳ. Evaluation and Critical Thinking

Based on your analysis of the two passages, discuss in groups and present your ideas on the following items.

1. What is the style of language in Text A, and why is or isn't it suitable for the issue under discussion?
2. Comment on the way the author presents Zeina Jabbour's experience with the IPK in Text A?
3. What do you think will be the consequences of the change in the weight of the IPK?
4. Why do data concerning other units of measurement appear in both passages in the discussion concerning the kilogram?
5. What are the advantages and disadvantages of the ways of determining weight in Text A and Text B?

Ⅴ. True/False Checking

Based on your understanding of the passages, decide whether each statement is true or false.

Text A

1. The IPK is cemented in a triple bell jar inside a temperature-and-humidity controlled vault.
2. Kilogram weights from countries all over the world can be brought in to gauge against the IPK by paying dues.
3. The Irish kilogram is weighed against three platinum-iridium working standards.
4. On the occasion of the formal verification of all kilogram prototypes, it was found that the 51 Meter Convention member states wanted the IPK.
5. It was unreasonable to depend on the safety of a manufactured item to determine mass.

Text B

6. The kilogram is a unit of light and energy.
7. The new method is more convenient for people to define a kilogram.
8. The cylinder which people used in the past would not give you a constant number when weighing things in different places.
9. In the past the kilogram is not fixed.

10. Energy is equal to the frequency of a photon multiplied by Planck's constant.

Ⅵ. Translation

Translate the following sentences into Chinese.

1. What I love best about the kilogram is its tangibility, its solid, sculpted form of shiny platinum and iridium. I'm referring to not just any kilogram but the quintessential one that resides here—the actual International Prototype Kilogram, or IPK, created in 1879 as the official standard of mass. (Para. 1, Text A)
2. Around him kilogram weights of various shapes and materials sit on colored plates under glass bell jars, like an assortment of fine cheeses. They have been delivered here from other countries to be reckoned in comparison with the IPK. (Para. 5, Text A)
3. Despite all the protective protocols and delicate procedures, it had mysteriously changed. No one can say whether the IPK has lost weight (perhaps by the gradual escape of gases trapped inside it from the start) or if most of the prototypes have gained (possibly by accumulating atmospheric contaminants). (Para. 11, Text A)
4. What's really fascinating here isn't that, practically speaking, the way most of us use the kilogram will change. It's how difficult it turned out to be to rigorously define a unit of mass at all. (Para. 5, Text B)
5. That enables scientists to define mass in terms of Planck's constant, an unchanging feature of the universe. An international coalition of science labs came together to make the most precise measurements of Planck's constant yet, certain to within just several parts per billion. (Para. 11, Text B)

Additional Reading

The First-Ever Permanently Magnetic Liquid

1. For the first time, scientists have created a permanently magnetic liquid. These liquid droplets can morph into various shapes and be externally manipulated to move around, according to a new study.
2. We typically imagine magnets as being solid, said senior author Thomas Russell, a distinguished professor of polymer science and engineering at the University of Massachusetts Amherst. But now we know that "we can make magnets that are liquid and they could conform to different shapes—and the shapes are really up to you".

3. The liquid droplets can change shape from a sphere to a cylinder to a pancake, he told Live Science. “We can [even] make it look like a sea urchin if we wanted.” [9 Cool Facts about Magnets]
4. Russell and his team created these liquid magnets by accident while experimenting with 3D printing liquids at the Lawrence Berkeley National Laboratory (where Russell is also a visiting faculty scientist). The goal was to create materials that are solid but have characteristics of liquids for various energy applications.
5. One day, postdoctoral student and lead author Xubo Liu noticed 3D-printed material, made from magnetized particles called iron-oxides, spinning around in unison on a magnetic stir plate. So when the team realized the entire construct, not just the particles, had become magnetic, they decided to investigate further.
6. Using a technique to 3D-printed liquids, the scientists created millimeter-size droplets from water, oil and iron-oxides. The liquid droplets keep their shape because some of the iron-oxide particles bind with surfactants—substances that reduce the surface tension of a liquid. The surfactants create a film around the liquid water, with some iron-oxide particles creating part of the filmy barrier, and the rest of the particles enclosed inside, Russell said.
7. The team then placed the millimeter-size droplets near a magnetic coil to magnetize them. But when they took the magnetic coil away, the droplets demonstrated an unseen behavior in liquids—they remained magnetized. (Magnetic liquids called ferrofluids do exist, but these liquids are only magnetized when in the presence of a magnetic field.)
8. When those droplets approached a magnetic field, the tiny iron-oxide particles all aligned in the same direction. And once they removed the magnetic field, the iron-oxide particles bound to the surfactant in the film were so jam-packed that they couldn’t move and so remained aligned. But those free-floating inside the droplet also remained aligned.
9. The scientists don’t fully understand how these particles hold onto the field, Russell said. Once they figure that out, there are many potential applications. For example, Russell imagines printing a cylinder with a non-magnetic middle and two magnetic caps. “The two ends would come together like a horseshoe magnet,” and be used as a mini “grabber”, he said.
10. In an even more bizarre application, imagine a mini liquid person—a smaller-scale version of the liquid T-1000 from the second *Terminator* movie—Russell said. Now imagine that parts of this mini liquid man are magnetized and parts aren’t. An external magnetic field could then force the little person to move its limbs like a marionette.
11. “For me, it sort of represents a sort of new state of magnetic materials,” Russell said. The findings were published on July 19 in the journal *Science*.

Unit Six Marine Engineering

Critical Reading

Text A

Naval Architecture (Excerpt)

Introduction

1. Naval architecture, the art and science of designing boats and ships to perform the missions and to meet the requirements laid down by the prospective owners and operators. It involves knowledge of mechanics, hydrostatics, hydrodynamics, steady and unsteady body motion, strength of materials, and design of structures.
2. A good naval architect and ship designer must have experience in a number of fields of engineering, as well as in the field of engineering economics. The architect must also understand the characteristics and properties of construction materials and be familiar with the latest and best methods of fabricating parts and joining them. Like other branches of engineering, naval architecture involves estimates and predictions of the final performance of the ship and all its parts, and of initial and operating costs. Such calculations must be made while the ship is still in the paper stage in the form of plans and specifications.

The Mission of a Ship

3. The detail requirements for any given ship are made up on the basis of its mission. Just how much cargo and how many passengers is it to carry? What are the requirements for manning the ship? What is to be its maximum or sustained speed, and under just what conditions? What must be its cruising radius, in terms of days as well as of distance? For a tug, the towing pull or free-running speed must be stated. For an icebreaker, capacity to push its way through ice of a specified thickness must be shown. For a warship, the armament must be given, and the weight and volume requirements for electronics equipment.

4. The wide variety of missions for watercraft produces a great number of distinct and specialized types. Considering naval architecture and design, these are subdivided roughly into two main classes: warships and merchant ships. The distinction is not always a sharp one. A naval transport may closely resemble a merchant passenger ship and may be designed in the same way. A fast motor cruiser may be designed like a PT boat without torpedoes, guns, and depth charges. In fact, the navy of any nation includes many merchant types, among them store and supply ships, oilers, ammunition-and-missile supply ships, repair ships, tenders for small craft, hospital ships, and personnel transports. The detail requirements for a specific ship can only be established by careful consideration of the system in which the ship is to operate: a transportation system in the case of a merchant ship or naval auxiliary and a ship-weapons system for a combatant vessel.
5. Commercial ship operators and designers are concerned with the overall door-to-door movement of general cargoes from inland point of origin to inland destination overseas. This concern has led to a growing trend toward unitization of cargo, that is, assembling of individual packages and cartons into larger units for greater ease of handling, for transferring from one transport mode to another, for less expensive packing, and for reduced opportunities for pilferage. Such concern results in problems for the naval architect regarding these large cargo units: should the cargo be strapped to pallets or stowed in containers? If containers are used, what size should be adopted for best efficiency and for coordination with land transportation? The ship designer must also be concerned with the layout of terminals and a choice between shore-based or ship-mounted cranes. The ship itself must be designed to function efficiently in the land-sea-terminal system.
6. In the transportation of bulk cargoes, either dry or liquid, economics has also introduced vital new problems for the naval architect. Cargo handling has not been involved in this equation, since efficient pumping methods and dry bulk handling techniques have long been available. The economic gains from increasing ship size pose problems in the design of terminals and offshore loading and discharge facilities. In addition, the necessity of transiting locks and canals places an upper limit on the size of some ships. The Panamax and Suezmax classes of tanker have been devised with the specific dimensions of the Panama Canal and Suez Canal, respectively, in mind.
7. For overall system efficiency, is it better to limit the size of the ships to suit the ports to be served or to transship from larger to smaller ships in a deepwater port? The naval architect must consider such problems before beginning the technical design of the ship itself.
8. In a similar manner, the naval designer must consider the system or systems in which the naval

ship is intended to operate. For example, an Ohio-class nuclear submarine is intended to be able to launch a Trident missile while submerged against any target on earth. An aircraft carrier, with its supporting and protecting screen of ships, is a system for projection of military force through airpower. Other ship-weapon systems have antisubmarine missions, either defensive or offensive. The effectiveness of all such ship systems is continually evaluated by the defense establishment in competition with other land-and-air based systems.

(820 words)

Glossary:

ammunition *n.* supply of bullets, bombs, grenades, etc. fired from weapons or thrown 军火

armament *n.* weaponry used by military or naval force 武器

cruise *v.* travel at a continuous speed 巡航

fabricate *v.* manufacture; make 制造

pilferage *n.* the act of stealing small amounts or small articles 偷窃

specifications *n.* a detailed description of design criteria for a piece of work 规范，说明

stow *v.* fill by packing tightly 填塞

tow *v.* drag behind 拖

torpedo *n.* armament consisting of a long cylindrical self-propelled underwater projectile that detonates on contact with a target 鱼雷

Text B

Ships in Waves

1. Considered as the environment for boats and ships of all kinds and sizes, the term sea is used to denote all waters large enough for the operation of these craft, from creeks and ponds to lakes and oceans. The wind and the ships moving across the sea create a pattern of undulations ranging from minute ripples to waves of gigantic size. The currents moving through it must also be taken into account in all ship operations and in some ship-design problems. The variations in density, resulting from the amount of salts in solution, determine the variable-ballast tank capacity of submarines and the ability of a submarine to "sit" on a layer of dense water while largely supported by a less dense layer above.
2. Considering the overall surface configuration, termed the seaway, the classical concept of a train of regular waves is highly unrealistic, but it has some practical uses. The normal seaway

is highly irregular, with waves of different heights and lengths traveling in many directions. For analytic purposes, it may be considered as made up of a multitude of very low waves, having a wide range of lengths and periods and traveling in various directions, superposed to produce the actual seaway. When this is done, a useful approach is to use statistical methods to define the seaway by its spectrum, which indicates the amplitudes of its many (theoretically infinite) wave components.

3. The sea is also home to teeming masses of marine life, many of which are detrimental to ships. Marine borers attack wood exposed on underwater portions of the hull. Barnacles cling to the underwater hull, roughening its surface and increasing the ship's resistance to travel through the water. Sea water is highly corrosive to most materials, and severe electrochemical effects cause rapid disintegration of submerged metals that are unprotected.

Ship Motions in Waves

4. Treated as a rigid body, a ship partakes of six oscillatory motions in a seaway. Three are translatory motions of the whole ship in one direction: (1) surge is the oscillation of the ship fore and aft; (2) sway is the motion from side to side; and (3) heave is the up-and-down motion. The other three oscillations are rotary; (4) roll is the angular rotation from side to side about a fore-and-aft axis; (5) pitch is the bow-up, bow-down motion about an athwartships axis; and (6) yaw is the swing of the ship about a vertical axis. Yawing is not necessarily oscillatory for every service condition. All six of these motions can and do take place simultaneously in a confused sea, so the situation is most complex.

5. The forces and moments caused by waves are balanced by three types of forces and moments opposing them: (1) those inertia reactions developed by the acceleration of the ship and cargo and the adjacent water; (2) those that result in damping the oscillatory ship motion or reducing its extent by the generation of surface gravity waves, eddies, vortexes, and turbulence, the energy required for setting up these disturbances is carried away and lost; (3) those of hydrostatic nature that act to restore the ship to a position of equilibrium as, for example, when the ship rolls to an angle greater than that called for by the exciting moment.

6. The behaviour of a ship in waves is too complex for the motions in all six degrees of freedom to be completely described mathematically. However, the longitudinal motions of pitching and heaving can be treated as a coupled system (neglecting surging), under the assumption that lateral motions do not exist at the same time or are reduced by stabilization to minimal values. Similarly, rolling can be treated along with heaving and swaying on the assumption that pitch

and heave do not occur or have negligible effect. Equations of motion can then be set up that equate the wave exciting forces and moments to the three types of forces associated with the motions that were described above.

7. The theory of rolling was developed in the 19th century by Froude. The theory of coupled pitching and heaving is more recent, stimulated by the work of Boris Korvin-Kroukovsky in the 1950s, who applied a so-called "strip" method in which the ship was divided longitudinally into strips or segments. The total force and moment acting on the ship and the resulting motions were assumed to be the result of the integration of all the forces in the individual strips without appreciable interference. Model tests in many laboratories have confirmed the basic soundness of this approach, although refinements are continually being made. Computer programs for solving the equations and calculating the pitch-heave motions of any ship are commonly used in the design stage.

8. The pioneering work of Manley St. Denis and Willard J. Pierson, Transactions of the Society of Naval Architects and Marine Engineers (1953), showed how the motions of a ship in an irregular seaway can be statistically described by assuming that the irregular motions are the sum of the ship's response to all the regular component waves of the seaway described by its spectrum. This powerful tool has permitted the extension of calculated motions (or those measured in a model tank) to the prediction of realistic irregular sea responses and hence to the comparative evaluation of alternative ship designs under realistic conditions.

9. Work by various investigators along the above lines has shown that longitudinal weight distribution and overall ship proportions have a much greater effect than details of hull form on pitching and heaving, and on the associated shipping of water, slamming, and high accelerations. In general, a short pitching period in relation to ship length is found to be advantageous in raising the limit of speed in rough head seas. This suggests concentration of heavy weights amidships, if possible, and favours long, slender hulls over short, squat ones.

Effect of Shape and Proportions

10. The most important single factor in cutting down the increased resistance, as well as motions, of ships running in waves appears to be a small fatness ratio; in other words, a small underwater volume compared with the ship length. This slenderness is difficult to work into ships intended to carry cargo but relatively easy for passenger ships. For reduction in the magnitude of ship motions in waves, it is important that the damping forces and moments be as large as practicable. Moderate flare in the above-water sections at bow and stern, large

beam compared with draft, and fineness of the underwater sections all help to achieve the result. A deep-sea sailing yacht embodies these characteristics to a high degree.

11. To keep the ship reasonably dry while undergoing the rolling, pitching, and heaving motions that remain, large freeboard is essential, especially at the bow. To prevent slamming under the bow when it lifts out of water and then drops heavily upon the surface, the forefoot underwater must also be deep.

12. A good degree of damping is most necessary to avoid deep rolling. If this cannot be achieved by a transverse form suited to the service, such as that of a sailing yacht with a deep fin keel, it is accomplished by adding long fins on each side in the form of roll-resisting or bilge keels. When placed along the lines of flow, these keels add little to the ship resistance in calm water.

13. Active roll-resisting fins serve to quench the greater part of the roll on a fast ship with a reasonable expenditure of weight, space, and cost. These fins, much shorter than bilge keels but extending several times as far outboard when in use, are rotated mechanically about transverse axes to produce angles of attack and girthwise forces which continually oppose the rolling motion. Since the moments of the roll-resisting forces increase as the square of the ship speed, the active fins are ineffective at low speeds.

14. Passive roll-resisting tanks of flume or U shape have been extensively installed in ships. In these, the tank dimensions and quantity of water or other liquid are arranged so that the liquid moves from one side of the ship to the other to counteract the rolling motion. Active tanks make use of controllable (and reversible) axial-flow propellers placed in ducts connecting the port and starboard tanks to control the flow.

15. Considering the vertical accelerations involved, pitching and heaving, or a combination of the two, are particularly objectionable for passenger comfort and safeguarding of cargo. They often necessitate a reduction in speed or a change of course. Some form of passive pitch-resisting fin may be evolved which will accomplish its primary purpose without introducing detrimental features.

(1,426 words)

Glossary:

equilibrium *n.* a stable situation in which forces cancel one another 平衡

inertia *n.* the tendency of a body to maintain its state of rest or uniform motion unless acted upon by an external force 惯性

longitudinal *adj.* running lengthwise 纵向的

magnitude *n.* the property of relative size or extent 大小

moment *n.* a turning force produced by an object acting at a distance (or a measure of that force) 力矩

oscillatory *adj.* having periodic vibrations 振动的

quench *v.* reduce the degree 按捺

spectrum *n.* a range of any of various kinds of waves 波谱

transverse *adj.* in a crosswise direction 横向的

undulation *n.* wavelike motion 波动

vertical *adj.* upright in position or posture 垂直的

Exercises

Ⅰ. Mindmap and Summary

Read the passages and finish the mindmap and summary exercises.

Text A

Analyze the supporting ideas of each part of the passage, and present them in the following mindmap.

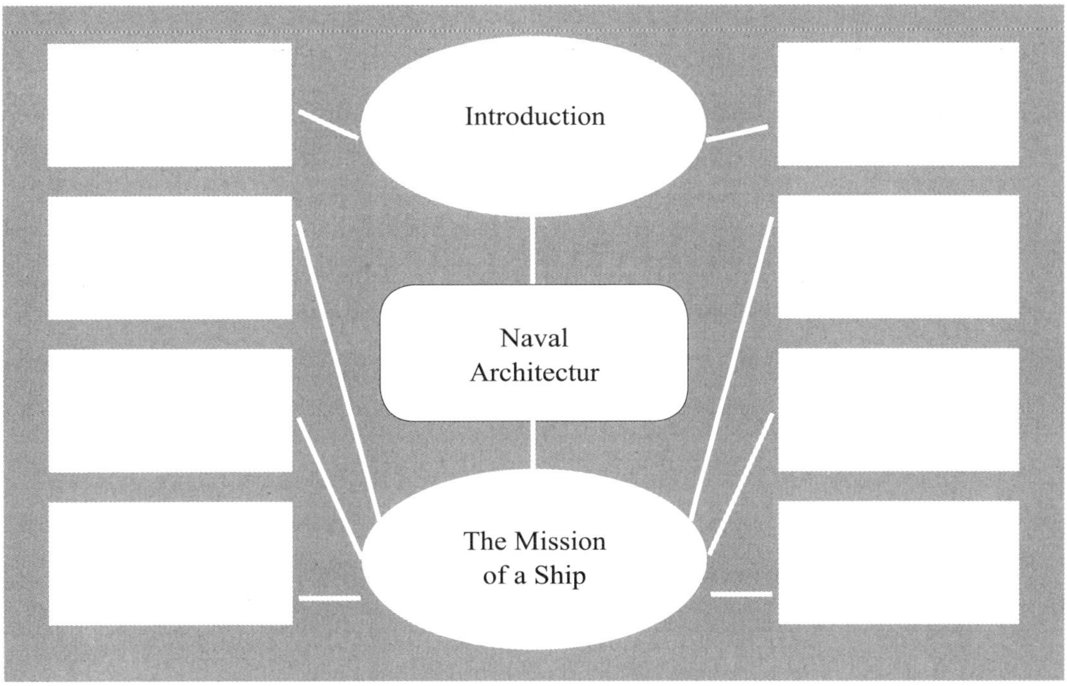

Text B

Analyze the information given in the passage, and write a summary.

____

____

____

____

____

____

____

____

____

Ⅱ. Paraphrasing

Interpret the following sentences in your own words.

1. The wide variety of missions for watercraft produces a great number of distinct and specialized types.
2. This concern has led to a growing trend toward unitization of cargo, that is, assembling of individual packages and cartons into larger units for greater ease of handling, for transferring from one transport mode to another, for less expensive packing, and for reduced opportunities for pilferage.
3. The total force and moment acting on the ship and the resulting motions were assumed to be the result of the integration of all the forces in the individual strips without appreciable interference.
4. Sea water is highly corrosive to most materials, and severe electrochemical effects cause rapid disintegration of submerged metals that are unprotected.
5. This slenderness is difficult to work into ships intended to carry cargo but relatively easy for passenger ships.

Ⅲ. Brief Answering

Answer the following questions based on information given in the two passages.

1. According to Text A, what is included in the expertise a naval architect must have?
2. Judged by design, what categories can watercrafts be classified into?

3. What is an important factor for military naval architect to take into consideration, as is presented in Text A?
4. What are the six oscillatory motions of a ship?
5. How are passive and active tanks different in resisting rolls?

Ⅳ. Evaluation and Critical Thinking

Based on your analysis of the two passages, discuss in groups and present your ideas on the following items.

1. Make a comment on the writing of the third paragraph of Text A.
2. Based on information given in Text A, what do you think are the differences between designing commercial ships and warships?
3. Compare the structure of "Ship Motions in Waves" and "Effect of Shape and Proportions" in Text B.
4. What aspects should also be considered when designing a ship except the ones presented in Text A and Text B?
5. Apart from ships, what are other types of marine architecture, and what is important for design them?

Ⅴ. True/False Checking

Based on your understanding of the passages, decide whether each statement is true or false.

Text A

1. A wide range of knowledge and experience is needed to be a good naval architect.
2. Ships for military and civilian use are mutually exclusive in type.
3. Certain problems ensue from unitization of cargo.
4. Size not only matters in designing ships, but also matters in designing terminals and related facilities.
5. Ship-weapon systems have defensive and offensive antisubmarine missions.

Text B

6. Ponds can be classified into the sea.
7. The normal seaway is made up of many very low waves.

8. Surge is a form of rotary oscillation.
9. Active roll-resisting fins function effectively at high and low speeds.
10. Active tanks work to make the liquid move from one side of the ship to the other to counteract the rolling motion.

Ⅵ. Translation

Translate the following sentences into Chinese.

1. The detail requirements for a specific ship can only be established by careful consideration of the system in which the ship is to operate: a transportation system in the case of a merchant ship or naval auxiliary and a ship-weapons system for a combatant vessel. (Para. 4, Text A)
2. In the transportation of bulk cargoes, either dry or liquid, economics has also introduced vital new problems for the naval architect. (Para. 6, Text A)
3. For overall system efficiency, is it better to limit the size of the ships to suit the ports to be served or to transship from larger to smaller ships in a deepwater port? The naval architect must consider such problems before beginning the technical design of the ship itself. (Para. 7, Text A)
4. For analytic purposes, it may be considered as made up of a multitude of very low waves, having a wide range of lengths and periods and traveling in various directions, superposed to produce the actual seaway. (Para. 2, Text B)
5. The total force and moment acting on the ship and the resulting motions were assumed to be the result of the integration of all the forces in the individual strips without appreciable interference. (Para. 7, Text B)

Additional Reading

China Tests World's 1st Robot Ship for Launching Small Rockets

1. China has built the world's first robotic, partially submersible boat for launching sounding rockets—a technology that will help meteorologists better understand the atmosphere over Earth's oceans.
2. Although the tests were conducted in 2016 and 2017, a paper describing the results of the first tests with the system has just been published. Traditionally, it's been difficult to study the three-quarters of the Earth's atmosphere that is found over water, because scientists have needed to

do so from planes or ships, both of which make for expensive expeditions. These endeavours are also usually more vulnerable to inclement weather compared to land-based observations.

3. That's where China's new boat comes in. Officially classified as an "unmanned semisubmersible vehicle", the new ship is designed to sail into bad weather, deploy a sounding rocket, and gather crucial data about the atmosphere and ocean.
4. Sounding rockets make brief flights through different layers of the atmosphere, in this case carrying meteorological equipment as high as 5 miles (8 kilometers) above the ocean.
5. "The unmanned semisubmersible vehicle is an ideal platform for marine meteorological environmental monitoring, and the atmospheric profile information provided by [sounding rockets] launched from this platform can improve the accuracy of numerical weather forecasts at sea and in coastal zones," co-author Jun Li, a researcher at the Institute of Atmospheric Physics of the Chinese Academy of Sciences, said in a statement accompanying the new paper.
6. Now, with these initial tests of the system complete, the team hopes to deploy a network of these boats, particularly in order to study typhoons, the equivalent of hurricanes in the western Pacific Ocean. They also hope to equip the boats themselves with more-advanced oceanography sensors, so the vessels can look down as well as up.
7. The test launches are described in a paper published Jan. 31 in the journal Advances in Atmospheric Science.

Unit Seven Nuclear Power

Critical Reading

Text A

Oil Is Out; Is Nuclear In?

1. Put yourself in an imaginary time machine and set the dial to around the year 2040. The exorbitant price of oil, now at $500 a barrel, has pushed a good chunk of the globe toward nuclear power.
2. If the world is on the cusp of such an era, will it result in weapons proliferation or accidents such as meltdowns? Or will a new way to harness energy emerge to compete with nuclear power?
3. Currently, more than 400 reactors in 30 countries supply about 16 percent of the world's electricity, according to the World Nuclear Association. Nuclear power is increasing steadily, with about 30 reactors under construction in 12 countries. Most reactors being ordered or planned are in Asia. And more than 40 developing countries have recently approached U.N. officials to express interest in starting nuclear power programs.
4. Nuclear energy advocates see the technology as a clean way to wean the world from expensive and environmentally damaging fossil fuels.
5. But for critics, the thought of a nuclearized world stirs controversy.
6. "The G8 countries can't even maintain adequate safety and security," said Paul Walker at Global Green USA, an environmental and arms control organization and affiliate of a group founded by former Soviet President Mikhail Gorbachev. "So the developing world presents a dangerous potential for terrorist attack or diversion of radioactive materials or major accident."
7. Experts are also concerned that the spread of nuclear power could spur weapons proliferation. Arab states could learn the wrong lessons from Iran, a nation that thumbs its nose at the West

while it allegedly builds a nuclear weapons program, said Charles Ferguson, senior fellow at the Council on Foreign Relations.

8. Neighbors of the Persian giant could follow its lead in sidestepping full disclosure of its nuclear program, or repeat its argument that unfriendly relations with the few Western countries that supply nuclear fuel have forced it to build it's own enrichment program, Ferguson said. Most countries that rely on nuclear power have only small programs without the ability to produce highly enriched uranium, which can be used in bomb making, he added.

9. "Iran thinks of itself as a top dog in the region, and that can appear threatening to Saudi Arabia and other Arab states," Ferguson said.

10. Surrounding nations could establish nuclear programs with the intent of supplying electricity but could later flirt with the idea of nuclear weapons, Ferguson said. Fear of a dominant Iran could prompt regional governments to keep their nuclear weapons options open, he added. Some could decide it's more cost effective to start enriching their own uranium.

11. "It's kind of a slow train wreck," Ferguson said.

12. Commercial reactors use low enriched uranium, which is non-weapons grade, although some research reactors use highly enriched uranium, Ferguson said. An enrichment facility can make low enriched uranium for commercial reactor fuel or highly enriched uranium for nuclear bombs, he said. "That's why there is such a big concern about the spread of enrichment technology," he said.

13. A turning point for the Middle East could come when its oil begins to dry up and the region starts clamoring for nuclear power.

14. "At that point Saudi Arabia will feel some economic pain," Ferguson said. "So they can legitimately say 'let's invest in nuclear' ."

15. To curtail moves from peaceful technology to weapons material production, experts recognize the need for strong regulatory bodies. But some authoritarian governments may not warm to the idea of answering to an independent government body, Ferguson said.

16. Ernie Moniz, director of Massachusetts Institute of Technology's energy initiative, said one way to curtail nuclear weapons proliferation is to push "fuel leasing", in which a country would receive a secure, fresh supply of low enriched uranium fuel and return the spent fuel to the supplier. The user would agree not to pursue further enrichment or reprocessing. Spent fuel can also be used to make weapons.

17. The United Arab Emirates is moving toward such an arrangement, although no country has formally adopted it, Moniz said.

18. The spread of nuclear power could also increase terrorists' access to nuclear materials, experts

said. Even now, some of the world's research reactors, which contain fuel suitable for bomb making, are lightly guarded and vulnerable to security breaches, Ferguson said.

19. Safety is another concern. There have been two major accidents—Chernobyl and Three Mile Island. Chernobyl was a result of major design deficiencies, violation of operating procedures and the absence of a safety culture, according to the World Nuclear Association.

20. "Chernobyl used an old design that could never have been licensed in the West," Moniz said. "I would expect that any future plant built anywhere will use modern safety features."

21. Regardless, it takes just one Chernobyl or Three Mile Island to spur panic, the fallout of which could halt nuclear power expansion worldwide, Ferguson said. The 1979 Three Mile Island accident raised many public questions about the safety and reliability of nuclear power and dealt a death blow to new plant construction in the United States, Walker said.

22. George Friedman, chief executive officer of Stratfor, a private intelligence firm, said nuclear weapons proliferation is unlikely. "The most remarkable thing is the lack of nuclear proliferation," he said. "In fact we've seen very little."

23. Nearly 200 states are party to the Nuclear Non Proliferation treaty, five of which have nuclear weapons. Other known nuclear powers are India and Pakistan. Iran and Syria are alleged to have nuclear weapons and Israel's nuclear weapons status remains unknown. Pakistani scientist A.Q. Khan is alleged to have shared nuclear secrets with DPRK, Libya and Iran, although he recently recanted his confession.

24. Simply creating a nuclear explosion is easy, Friedman said. "But creating a nuclear weapon is fiendishly difficult," he said, "explaining why weapons proliferation fears are unfounded."

25. Iran, for example, does not have the advanced electronics or other engineering capabilities to deliver such an explosion, he said. "Eighty percent of a weapons program has nothing to do with nuclear weapons."

26. But he does expect increased civilian use of nuclear power in a future with a dwindling supply of affordable fossil fuels.

27. One problem is that the electricity that nuclear plants produce cannot be stored well. An airplane, for example, cannot be flown on batteries, Friedman said. "And you can't drive a truck using electricity," he added. Building a reactor can also take years.

28. "So the real issue is going to be storage," Friedman said.

29. Friedman expects Japan, which imports 100 percent of its oil, to move increasingly toward nuclear power. Other countries would likely include industrial nations with limited domestic resources like South Korea, he said. European countries would also fall into the mix, at least

until a better technology comes along.

30. But nuclear power's limitations could spur a drive toward other technologies.

31. "If I were to bet, I would bet on space-based solar power," Friedman said.

32. Direct conversion of solar energy on earth is inefficient, mostly because of the amount of land it uses, Friedman said. But beaming microwave energy to earth would be more viable later in the century. This would especially be the case if the military's use of space soars, which would eventually cause launch costs to decline, he said.

33. In line with such predictions is a 2007 report by the Pentagon's National Security Space Office entitled Space-Based Solar Power as an Opportunity for Strategic Security, which encourages the government to start developing space power.

34. As for which energy source will reign later in the century, Friedman expects a range of competing ideas but says that one dominant one will emerge, just as hydrocarbons did in the 19th century.

35. "I expect ferment, crisis and then a single solution (will) emerge," he said.

(1,312 words)

Glossary:

affiliate *n.* a subsidiary or subordinate organization that is affiliated with another organization 附属机构

breach *n.* a failure to perform some promised act or obligation 违背

curtail *v.* place restrictions on 限制

exorbitant *adj.* greatly exceeding bounds of reason or moderation 过高的

ferment *n.* a state of agitation or turbulent change or development 发酵

meltdown *n.* severe overheating of the core of a nuclear reactor resulting in the core melting and radiation escaping 核心熔融

proliferation *n.* a rapid increase in number (especially a rapid increase in the number of deadly weapons) 增殖（尤指：武器扩散）

recant *v.* formally reject or disavow a formerly held belief, usually under pressure 放弃主张

viable *adj.* capable of being done with means at hand and circumstances as they are 可行的

Text B

The Necessity of Nuclear Power

1. Nuclear energy has been a controversial topic ever since the first reactor powered four 200-watt light bulbs in the Idaho desert in 1951. Today, the United States is gearing up for the next generation of reactor designs, breathing new life into the decades-old debate. Here, Patrick Moore and Anna Aurilio present compelling arguments both for and against pursuing nuclear power as an answer to the country's energy problems.

Nuclear Energy Provides Practical Baseload Power by Patrick Moore

2. When I helped found Greenpeace in the 1970s, my colleagues and I were firmly opposed to nuclear energy. But times have changed. I now realize nuclear energy is the only non-greenhouse gas-emitting power source that can effectively replace fossil fuels and satisfy growing demand for energy.
3. Nuclear power plants are a practical option for producing clean, cost-effective, reliable and safe baseload power.
4. Nuclear energy is affordable. The average cost of producing nuclear energy in the United States is less than two cents per kilowatt-hour, comparable with coal and hydroelectric.
5. Nuclear energy is safe. In 1979, a partial reactor core meltdown at Three Mile Island frightened the country. At the time, no one noticed Three Mile Island was a success story; the concrete containment structure prevented radiation from escaping into the environment. There was no injury or death among the public or nuclear workers. This was the only serious accident in the history of nuclear energy generation in the United States. Today, 103 nuclear reactors quietly deliver 20 percent of America's electricity.
6. Spent nuclear fuel is not waste. Recycling spent fuel, which still contains 95 percent of its original energy, will greatly reduce the need for treatment and disposal.
7. Nuclear power plants are not vulnerable to terrorist attack. The five-feet-thick reinforced concrete containment vessel protects contents from the outside as well as the inside. Even if a jumbo jet did crash into a reactor and breach the containment, the reactor would not explode.
8. Nuclear weapons are no longer inextricably linked to nuclear power plants. Centrifuge technology now allows nations to produce weapons-grade plutonium without first constructing

a nuclear reactor. Iran's nuclear weapons threat, for instance, is completely distinct from peaceful nuclear energy generation, as they do not yet possess a nuclear reactor.

9. New technologies, such as the reprocessing system recently introduced in Japan (in which the plutonium is never separated from the uranium) can make it much more difficult to manufacture weapons using civilian materials.

10. Finally, excess heat from nuclear reactors offers a practical path to the "hydrogen economy", and can address the increasing shortage of fresh water through desalinization.

11. A combination of nuclear energy, wind, geothermal and hydro is the most environmentally-friendly way to meet the world's increasing energy needs. Nuclear power plants can play a key role in producing safe, clean, reliable baseload electricity.

12. An advisor to government and industry, Dr. Patrick Moore is a co-founder and former leader of Greenpeace, and chair and chief scientist of Greenspirit Strategies Ltd. in Vancouver, Canada. He and former Environmental Protection Agency administrator Christine Todd Whitman are co-chairs of the Clean and Safe Energy Coalition, which supports increased use of nuclear energy.

Nuclear Energy is Simply Not Necessary by Anna Aurilio

13. Nuclear energy is too expensive, too dangerous, and too polluting. And despite claims from the nuclear industry, it's simply not necessary either for our future electricity needs or to meet the very real challenge of global warming. Worldwide, renewable alternatives such as wind, solar and geothermal power, along with small decentralized heat and power cogeneration plants, already produced 92 percent as much electricity as nuclear power did in 2004 and those sources are growing almost six times faster. A recent study prepared by Synapse Energy Economics found that by using clean energy technologies in the next twenty years, the U.S. could cut our reliance on nuclear in half, reduce projected carbon dioxide emissions from electricity by 47% and save consumers $36 billion annually.

14. After 50 years and more than $150 billion dollars in subsidies, the nuclear industry is still unable to build a plant on its own. With the new incentives in the 2005 Energy Policy Act, taxpayers would be covering 60 to 90 percent of the generation cost of electricity from a new nuclear plant. What do we get for our money?

15. In a post-9/11 world, nuclear facilities will always be a tempting target for terrorists, and government studies have highlighted the weaknesses in our current safeguards.

16. Even without attackers, the danger of an accident is ever-present. The Davis-Besse plant in

Ohio narrowly avoided a disaster in 2002 when inspectors found a hole that had corroded almost all the way through a pressure vessel, leaving just 3/16 of an inch of steel preventing the release of radioactive steam. Instead of clamping down, the Nuclear Regulatory Commission seems more intent on loosening safety rules to help aging plants keep operating for longer.

17. And when plants are operating perfectly, they're still producing high-level radioactive waste. No country in the world has solved the problem of how to dispose of it, and even the most optimistic advanced reactor designs will continue adding to the lethal mountain of waste already produced.

18. The argument that nuclear energy is our best bet to reduce global warming emissions only makes sense if you pretend that coal is the only other option. That's a false choice, and it ignores the rapidly developing range of energy efficiency and clean, renewable energy sources. Whatever challenges still face technologies like solar and wind power, they pale compared to the fundamental security and environmental problems that won't be fixed by any new reactor design. For 30 years, no one has ordered or built a new nuclear plant, for very good economic reasons. Now Congress and the nuclear industry are trying to distort the market with new subsidies. They're pushing a technology with serious health, safety and economic risks, and in doing so diverting research dollars away from better alternatives.

19. Anna Aurilio is the Legislative Director for U.S. PIRG responsible for policy development, research and advocacy on energy issues and anti-environmental subsidies. She has testified numerous times before House and Senate Science, Energy and Appropriations committees. Ms. Aurilio received a bachelor's degree in Physics from the University of Massachusetts at Amherst in 1986 and a Master's degree in Environmental Engineering from the Massachusetts Institute of Technology in 1992. Prior to receiving her Master's degree, Ms. Aurilio was a Staff Scientist with the National Environmental Law Center, and the PIRGs' National Litigation Project for three years.

(1,017 words)

Glossary:

baseload *n.* the minimum amount of power that a utility or distribution company must make available to its customers, or the amount of power required to meet minimum demands based on reasonable expectations of customer requirements 基底负载

centrifuge *n.* separating substances by whirling them 离心分离

cogeneration *n.* the use of a heat engine or a power station to simultaneously generate both electricity and useful heat 热电联供

corrode *v.* cause to deteriorate due to the action of water, air, or an acid 腐蚀

desalinization *n.* the removal of salt (especially from sea water) 脱盐作用

geothermal *adj.* of or relating to the heat in the interior of the earth 地热的

jumbo *adj.* of great mass; huge and bulky 大型的

plutonium *n.* a solid silvery grey radioactive transuranic element whose atoms can be split when bombarded with neutrons; found in minute quantities in uranium ores but is usually synthesized in nuclear reactors; 13 isotopes are known with the most important being plutonium 239 钚（放射性元素）

Exercises

Ⅰ. Outline and Mindmap

Read the passages and finish the outline and mindmap exercises.

Text A

Analyze the main idea of each part and find out how the writer organizes this passage. Figure out the logical connection between each part, and work out how coherence and cohesion are achieved.

Para. 1 ~ 2	A scenario of the future concerning nuclear power
Para. 3 ~ 5	(1)____________________
Para. 6 ~ 11	Worries about weapons proliferation
Para. 12 ~ 14	(2) __________ of weapons proliferation
Para. 15 ~ 23	Efforts to (3) __________ nuclear weapons
Para. 24 ~ 35	(4) __________ is less likely
Para. 31-36	Alternatives to nuclear energy and the expected future

Text B

Analyze and classify the arguments given by the two experts concerning whether it is sensible to adopt nuclear energy, compare arguments under each category and present them in the following mindmap. One category has been given.

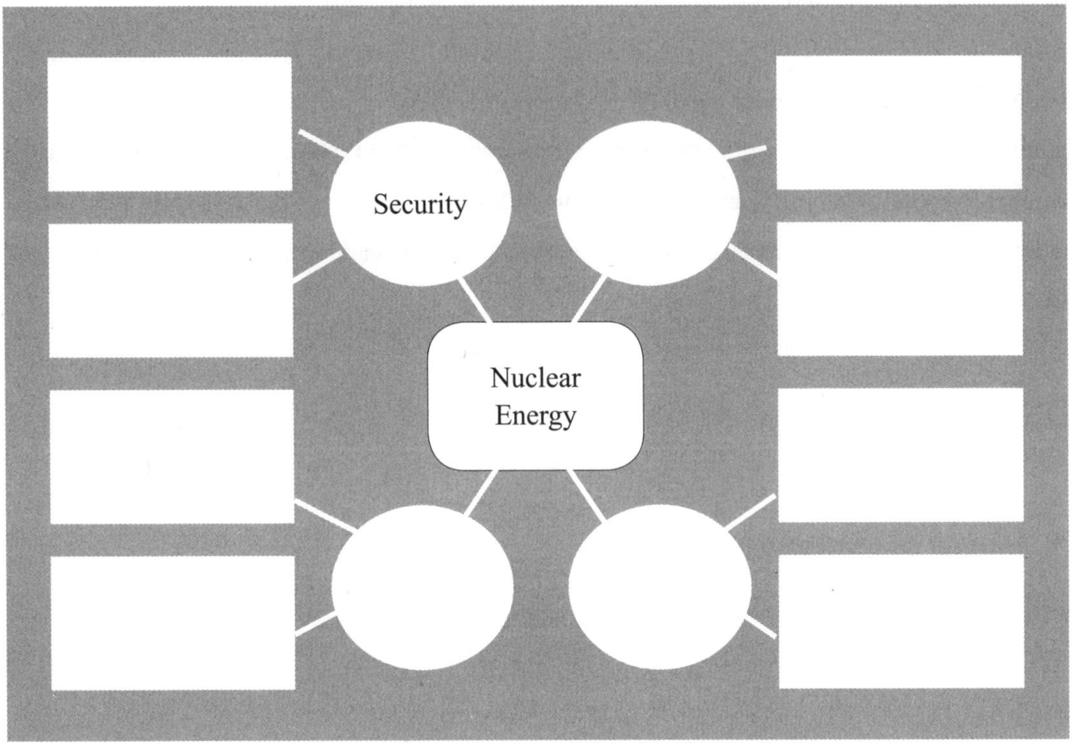

Ⅱ. Paraphrasing

Interpret the following sentences in your own words.

1. The exorbitant price of oil, now at $500 a barrel, has pushed a good chunk of the globe toward nuclear power.
2. Experts are also concerned that the spread of nuclear power could spur weapons proliferation. Arab states could learn the wrong lessons from Iran, a nation that thumbs its nose at the West while it allegedly builds a nuclear weapons program.
3. Simply creating a nuclear explosion is easy, Friedman said. "But creating a nuclear weapon is fiendishly difficult," he said, explaining why weapons proliferation fears are unfounded.
4. Finally, excess heat from nuclear reactors offers a practical path to the "hydrogen economy", and can address the increasing shortage of fresh water through desalinization.
5. Whatever challenges still face technologies like solar and wind power, they pale compared to the fundamental security and environmental problems that won't be fixed by any new

reactor design.

Ⅲ. Brief Answering

Answer the following questions based on information given in the two passages.

1. What are the mainstream ideas held by people concerning nuclear energy as presented in Text A?
2. In Text A, what kind of problem is presented by mentioning the example of Iran? To thwart such a problem, what measures should be taken?
3. According to Text A, what is the difference between civilian usage and non-civilian usage of nuclear energy?
4. How did people respond to the limitations of nuclear energy?
5. How do you understand "nuclear energy provides practical baseload power"?
6. How do Anna Aurilio and Patrick Moore differ in attitudes towards the usage of nuclear energy? What aspects are probed into in their argument?

Ⅳ. Evaluation and Critical Thinking

Based on your analysis of the two passages, discuss in groups and present your ideas on the following items.

1. Based on information presented in Text A and Text B, what are the threats posed by the use of nuclear energy? How to tackle them?
2. In dealing with issues concerning weapons proliferation, should there be a party that has the final say? If there should be one, who should it be? If not, how would such issues be addressed?
3. Examine all the aspects presented by the two parties in Text B, then discuss each of them, and decide which is most convincing? If you find neither side provides satisfactory argument concerning a certain point, demonstrate your opinion on it?
4. What are the alternatives to nuclear energy, and in your opinion what is the most environment-friendly way to meet the ever-increasing need of energy in the world?
5. Summarize Anna Aurilio and Patrick Moore's proofs in their argument. What is the purpose of presenting the ideas of both sides in Text B?

Ⅴ. True/False Checking

Based on your understanding of the passages, decide whether each statement is true or false.

Text A

1. Nuclear energy provides a cleaner way to address energy shortage than fossil fuels.
2. The spread of nuclear power is sure to spur weapons proliferation.
3. Surrounding nations of Iran would establish nuclear programs to supply electricity as much as create nuclear weapons.
4. The best way to prevent moves from peaceful technology to weapons material production is to establish an independent government body.
5. Cost of solar energy would decrease with the increase in the military's use of space.

Text B

6. Spent nuclear fuel requires less for treatment and disposal.
7. Nuclear reactors are free from terrorist attack.
8. It is not necessary to construct a nuclear reactor to produce weapons-grade plutonium.
9. Solar and wind power face greater challenges when compared to any new reactor design.
10. Advanced reactor designs would resolve the problem of continuing adding to the lethal mountain of nuclear waste.

Ⅵ. Translation

Translate the following sentences into Chinese.

1. Experts are also concerned that the spread of nuclear power could spur weapons proliferation. Arab states could learn the wrong lessons from Iran, a nation that thumbs its nose at the West while it allegedly builds a nuclear weapons program, said Charles Ferguson, senior fellow at the Council on Foreign Relations. (Para. 7, Text A)
2. But he does expect increased civilian use of nuclear power in a future with a dwindling supply of affordable fossil fuels. (Para. 26, Text A)
3. One problem is that the electricity that nuclear plants produce cannot be stored well. An airplane, for example, cannot be flown on batteries, Friedman said. "And you can't drive a truck using electricity," he added. Building a reactor can also take years. (Para. 27, Text A)
4. Nuclear weapons are no longer inextricably linked to nuclear power plants. Centrifuge technology now allows nations to produce weapons-grade plutonium without first constructing a nuclear reactor. Iran's nuclear weapons threat, for instance, is completely distinct from peaceful nuclear energy generation, as they do not yet possess a nuclear reactor. (Para. 8, Text B)

5. Whatever challenges still face technologies like solar and wind power, they pale compared to the fundamental security and environmental problems that won't be fixed by any new reactor design. (Para. 18, Text B)

Additional Reading

China Achieves Major Breakthrough in Nuclear Energy

1. Chinese researchers recently reported a breakthrough in increasing the usage rate of uranium materials from current 1 percent to more than 95 percent.
2. Xu Hushan, deputy head of the Institute of Modern Physics under Chinese Academy of Sciences, released the latest development at a news conference on Thursday.
3. Xu said that researchers at the institute carried out laboratory simulation to prove the effectiveness of a new system, called Accelerator Driven Advanced Nuclear Energy System (ADANES).
4. The new system would reduce the nuclear waste to less than 4 percent of the spent fuel, lowering the radiation lifetime from hundreds of thousands of years to about 500 years.
5. China has launched its strategy to develop clean, efficient, safe and reliable nuclear energy to secure the country's economic and social sustainable development, but problems exist in usage rate and waste disposal, Xu said.
6. In 2011, the Chinese Academy of Sciences launched the program of developing future accelerator driven subcritical transmutation system (ADS), but Chinese researchers later realized that the ADS system had little economic competitiveness and huge technological challenges.
7. Chinese researchers raised the new ADANES system concept and made the technological breakthrough in six years to make nuclear fission become a sustainable, safe and clean energy source.
8. China's "Artificial sun" device set is to be commissioned in 2020.
9. The photo shows the Experimental Advanced Superconducting Tokamak in Hefei, East China's Anhui province, which is dubbed as "artificial sun", Aug 16, 2018.
10. CHENGDU—The HL-2M Tokamak, China's next-generation "artificial sun", is expected to be operational in 2020 as installation work has gone smoothly since the delivery of the coil

system in June.

11. Designed to replicate the natural reactions that occur in the sun using hydrogen and deuterium gases as fuels, the device aims at providing clean energy through controlled nuclear fusion.

12. The new apparatus, with a more advanced structure and control mode, is expected to generate plasmas hotter than 200 million degrees Celsius, said Duan Xuru, head of the Southwestern Institute of Physics under the China National Nuclear Corporation.

13. Duan was quoted at the ongoing 2019 China Fusion Energy Conference held in Leshan, southwest China's Sichuan Province.

14. The artificial sun will provide key technical support for China's participation in the International Thermonuclear Experimental Reactor project, as well as the self-designing and building of fusion reactors, he noted.

Unit Eight Genetic Engineering

Critical Reading

Text A

What Are GMOs and GM Foods?

1. A genetically modified organism, or GMO, is an organism that has had its DNA altered or modified in some way through genetic engineering.
2. In most cases, GMOs have been altered with DNA from another organism, be it a bacterium, plant, virus or animal; these organisms are sometimes referred to as "transgenic" organisms. Genetics from a spider that helps the arachnid produce silk, for example, could be inserted into the DNA of an ordinary goat.
3. It sounds far-fetched, but that is the exact process used to breed goats that produce silk proteins in their goat milk, Science Nation reported. Their milk is then harvested, and the silk protein is then isolated to make a lightweight, ultrastrong silk material with a wide range of industrial and medical uses.
4. The dizzying range of GMO categories is enough to boggle the mind. CRISPR, a novel genome editing tool, has allowed geneticists to breed GMO pigs that glow in the dark by inserting jellyfish bioluminescence genetic code into pig DNA. CRISPR is opening doors to genetic modifications the likes of which were unimaginable just a decade ago.
5. These are more comparatively wild examples, but GMOs are already very common in the farming industry. The most common genetic modifications are designed to create higher yield crops, more consistent products, and resist pests, pesticides and fertilizer.

Genetically Modified Food

6. According to the National Library of Medicine (part of the National Center for Biotechnology Information, NCBI), genetically engineered, or genetically modified(GM), foods are those that

have had foreign genes from other plants or animals inserted into their genetic codes. This has resulted in foods that are consistently flavored, as well as resistant to disease and drought.

7. However, the NCBI also maintains a list of potential risks associated with GM foods, including genetic alterations that can cause environmental harm. Specifically, it's possible that modified organisms could be inbred with natural organisms, leading to the possible extinction of the original organism. For instance, the banana tree is propagated entirely through cloning methods. The bananas themselves are sterile.

8. By far, the biggest use of GMO technology is in large-scale agricultural crops. At least 90% of the soy, cotton, canola, corn and sugar beets sold in the United States have been genetically engineered. The adoption of herbicide-resistant corn, which had been slower in previous years, has accelerated, reaching 89% of U.S. corn acreage in 2014 and 2015, according to the U.S. Department of Agriculture.

9. One of the biggest draws for widespread adoption of GMO crops is pest resistance. According to the World Health Organization, one of the most widely used methods for incorporating pest resistance into plants is through Bacillusthuringiensis (Bt) genetics, a bacterium that produces proteins that repel insects. GMO crops that are modified with the Bt gene have a proven resistance to insect pests, thus reducing the need for wide-scale spraying of synthetic pesticides.

Are GMOs Safe?

10. Anti-GMO activists argue that GMOs can cause environmental damage and health problems for consumers.

11. One such anti-GMO organization is the Center for Food Safety, which calls the genetic engineering of plants and animals potentially "one of the greatest and most intractable environmental challenges of the 21st century".

12. "Genetically modified foods have been linked to toxic and allergic reactions, sickness, sterile and dead livestock, and damage to virtually every organ studied in lab animals," according to the Institute for Responsible Technology, a group of anti-GMO activists.

13. "Most developed nations do not consider GMOs to be safe," according to the Non-GMO Project. "In more than 60 countries around the world, including Australia, Japan and all of the countries in the European Union, there are significant restrictions or outright bans on the production and sale of GMOs."

14. As You Sow is a nonprofit environmental watchdog focusing its research on how corporate

actions affect our environment, including food production. According to Christy Spees, a program manager with As You Sow, GMO foods are dangerous "because the modifications are centered around resistance to toxic substances, such as pesticides and certain fertilizers. When dangerous chemicals are applied, plants use them to grow, and the food itself can be detrimental to our health".

Why GMOs Are Good

15. Many scientific organizations and industry groups agree that the fear-mongering that runs through discussions of GMO foods is more emotional than factual. "Indeed, the science is quite clear: crop improvement by the modern molecular techniques of biotechnology is safe," the American Association for the Advancement of Science (AAAS) said in a 2012 statement.
16. "The World Health Organization, the American Medical Association, the U.S. National Academy of Sciences, the British Royal Society, and every other respected organization that has examined the evidence has come to the same conclusion: Consuming foods containing ingredients derived from GM crops is no riskier than consuming the same foods containing ingredients from crop plants modified by conventional plant improvement techniques," according to the AAAS.
17. Others point to the benefits of sturdier crops with higher yields. "GM crops can improve yields for farmers, reduce draws on natural resources and fossil fuels and provide nutritional benefits," according to a statement on the website for Monsanto, the world's largest manufacturer of GMOs.
18. Monsanto and other agriculture companies have a financial stake in the research and messaging surrounding GM foods and have the resources to fund research that reinforces their narrative. However, although there are plenty of scientific data that demonstrates the safety, efficacy and resilience of GM crops, genetic modification remains a comparatively new scientific field.

GMO Labeling Debate

19. The argument over the development and marketing of GMO foods has become a political hot potato in recent years.
20. In November 2015, the FDA issued a ruling that only requires additional labeling of foods derived from genetically engineered sources if there is a material difference—such as a

different nutritional profile—between the GMO product and its non-GMO equivalent. The agency also approved Aquaadvantage Salmon, a salmon designed to grow faster than non-GMO salmon.

21. According to Monsanto, "there is no scientific justification for special labeling of foods that contain GM ingredients. We support these positions and the FDA's approach."

22. According to GMO Answers, an industry group comprised of Monsanto, DuPont, Dow Agrosciences, Bayer, BASF, Cropscience and Syngenta, GMO agricultural products are "by far the most regulated and tested product in agricultural history".

23. Additionally, their website states that "many independent scientists and organizations around the world—such as the U.S. National Academy of Sciences, United Nations Food and Agriculture Organization, World Health Organization, American Medical Association and the American Association for the Advancement of Science—have looked at thousands of scientific studies and concluded that GM food crops do not pose more risks to people, animals or the environment than any other foods".

24. The political issue that GMOs have become is almost as conductive as the scientific debate. However, after much discussion among various lawmakers across the U.S., the National Bioengineered Food Disclosure Standard (NBFDS)was passed into law at the beginning of 2019.

25. According to the NBFDS current federal statutes, starting in 2020, all food must bear a bioengineered (BE)label if it contains more than 5% bioengineered material. States are free to impose their own labeling requirements as well, though it seems that most jurisdictions are waiting for federal laws to be implemented before working on new legislation. One thing is for certain: the scientific and political discussions surrounding GMO foods aren't going away any time soon.

(1,232 words)

Glossary:

arachnid *n.* air-breathing arthropods characterized by simple eyes and four pairs of legs 蛛形纲动物

detrimental *adj.* harmful 有害的

incorporate *v.* make into a whole 合并

intractable *adj.* difficult to manage 棘手的

isolate *v.* place or set apart 隔离

monger *v.* sell or offer for sale from place to place 兜售

propagate *v.* reproduce 繁殖

reinforce *v.* strengthen 增强

resilience *n.* the ability to return to its usual state 适应力

Text B

Mosquito and Cucumber Salad Anyone?

1. Could a genetic hybrid of a mosquito and a sea cucumber spell the end of malaria one of the World's most deadly diseases? Do you live in a developed country and feel unconcerned about malaria? Perhaps you should think again. With the explosion of easy international travel, imported cases of malaria are reported more frequently. And the emergence of drug-resistant strains means the disease is appearing again in areas where it was previously under control.
2. There has never been a greater need for innovative preventative measures and new anti-malarial drugs. But have we found an unlikely ally in a gelatinous blob from the bottom of the Ocean.

Malaria's Deadly Toll

3. Malaria is an infectious disease caused by the plasmodium parasite that's transmitted by mosquitoes. Although malaria is largely a preventable and treatable disease, one person dies of malaria every 30 seconds. It rivals HIV and tuberculosis as the world's most deadly infection and the majority of its victims are under five years old.
4. Malaria causes severe illness in 500 million people worldwide each year, and kills more than one million. It is estimated that 40% of the world's population are at risk. Malaria transmission occurs primarily in large areas of Central and South America, sub-Saharan Africa, the Indian subcontinent, Southeast Asia and the Middle East.
5. In addition to an impact on individual health, malaria also has a significant socioeconomic impact. The disease causes an average loss of 1.3% annual economic growth in countries with a high incidence of infection. Furthermore, malaria has lifelong effects through increased poverty, impaired learning and decreased attendance in schools and the workplace.

How it All Begins—the Malaria Parasite Life Cycle

6. The transmission of the parasite plasmodium begins with the female mosquito, which needs blood to make eggs. When the mosquito bites an infected human it ingests parasite sperm and

eggs. These then unite in the stomach of the mosquito to form what's known as ookinetes-cells that become embedded in the stomach wall.

7. These ookinetes migrate through the mosquito's stomach wall and produce thousands of infectious daughter cells known as sporozoites. After 10—20 days they move to the mosquito salivary glands and are ready to infect another human.

8. Once inside the human body, the sporozoites are carried by the blood to the victim's liver where they hide from the immune system. In the liver they invade the cells and multiply into thousands of cells. After 9—16 days they return to the blood and infiltrate red blood cells where again they stay invisible to immune surveillance. Within the red blood cells they multiply once more forming parasite sperm and egg cells. This process destroy's the victim's red blood cells which releases the parasitic cells into the bloodstream to be ingested by a mosquito thus renewing the transmission cycle.

9. Patients with malaria exhibit extreme feverish attacks, flu-like symptoms, tiredness, diarrhea, nausea, vomiting and shivering while the malarial parasite damages the liver and blood cells. The severity of the symptoms depends on several factors, such as the species (type) of infecting parasite and the patient's acquired immunity and genetic background.

Cucumbers that Eat themselves Alive—and Malaria

10. The malaria parasite plasmodium has been studied for decades, but because of its ability to evade the immune system a vaccine has been hard to develop.

11. Most drugs that are available are active against the parasites while they are in the human blood. However, there is an emergence of parasite strains that are resistant to many of the existing anti-malarial drugs thus making the discovery of new therapies essential.

12. Recently, an International team of researchers, led by Professor Bob Sinden from the Imperial College in London, has found that the sea cucumber makes a chemical that is toxic to the fatal malarial parasite.

13. The sea cucumber is a worm-like scavenger that feeds on plankton and debris in the ocean and is made of a tough gelatinous tissue. A slimy mass of muscular tissue, the cucumber has the unusual ability to violently expel its internal body organs during times of stress. These can later be regenerated. Another fascinating feature is their ability to slowly digest themselves to cope with periods of starvation.

Cucumbers that Cure

14. Some varieties of sea cucumbers are said to have excellent healing properties so it may be no surprise that this creature has found medicinal use, both formally and in alternative therapies. Extracts have found their way into oils, creams and cosmetics. Sea cucumber extract has also been shown to heal wounds more quickly and reduce scarring. Other extracts from sea cucumbers inhibit blood vessel formation thus making it an excellent potential therapy for cancer.
15. Researchers have now found that this slug like creature also produces a protein which impairs development of the malaria parasite. Called lectin (CEL-Ⅲ), this protein has been shown to damage human and rat red blood cells. Lectin is poisonous to the parasites when they are still in the early ookinete stage of development—before they migrate out of the mosquito's stomach to produce the millions of sporozoites which end up in the insect's saliva.
16. Inhibiting this ookinete development is a potential way to eliminate the transmission of sporozoites to the human host thus breaking the transmission cycle.

Part Cucumber and Part Fly

17. In order to inhibit ookinete development within the stomach, the research team has developed genetically modified mosquitoes. They are just like regular mosquitoes accept for one gene. This newly introduced gene is part lectin-making gene from the sea cucumber and part gene from the mosquito that makes a substance that is released into the stomach during feeding. In this way the toxin can be introduced into the stomach where it can kill the ookinete cells.

Still No Solution (of should we say "dressing")

18. The concept of genetically modifying mosquitoes to disrupt parasite development prior to human infection is a novel approach to control the disease that is gaining recognition due to limited success elsewhere.
19. Currently, the only treatments against the parasite once they are in human blood are chloroquine, quinine, doxycycline, mefloquine and a few others. Unfortunately, since the most afflicted countries tend to be the poorest there has been little commercial incentive to develop new drugs. But with the disease now on our doorsteps perhaps activity here

will step up.

20. The rapid spread of antimalarial drug resistance over the past few decades has required more intensive monitoring to make sure there is proper management of clinical cases and early detection of changing patterns of resistance. Countries are being assisted in strengthening their drug resistance surveillance systems.

21. Mosquitoes have also built up resistance to insecticides which have been used in abundance.

22. But the deployment of super mossies that can fight off infection, like the cucumber salad, faces considerable challenges. In order to carry out this endeavor, thousands of mosquito embryos must be injected with a gene encoding for the anti-malarial protein and then the genetically modified adult mosquitoes released into the wild. And, unfortunately the toxic protein does not totally remove all parasites from all mosquitoes and as such, at this stage of development, would not be effective enough to prevent transmission of malaria to humans.

23. Furthermore, the genetically modified version of the mosquito would have to become the predominant species which is very difficult to achieve.

24. More research needs to be done but this finding is a first step toward developing a new method of preventing transmission of malaria thanks to the lowly sea cucumber.

(1,275 words)

Glossary:

impair *v.* make worse or less effective 损害

infectious *adj.* caused by infection or capable of causing infection 传染的

infiltrate *v.* enter by penetrating the interstices 潜入

ingest *v.* serve oneself to, or consume regularly 摄取

parasite *n.* an animal or plant that lives in or on a host to obtain nourishment from it without benefiting or killing the host 寄生虫

salivary *adj.* of or relating to saliva 唾液的

scavenger *n.* any animal that feeds on refuse and other decaying organic matter 清道夫

surveillance *n.* close observation of a person or group (usually by the police) 监视， 监督

toxic *adj.* poisonous 有毒的

Exercises

Ⅰ. Mindmap and Summary

Read the passages and finish the mindmap and summary exercises.

Text A

Analyze the argument presented in the passage concerning people's attitudes toward the GMOs, and complete the following mindmap. One category has been given.

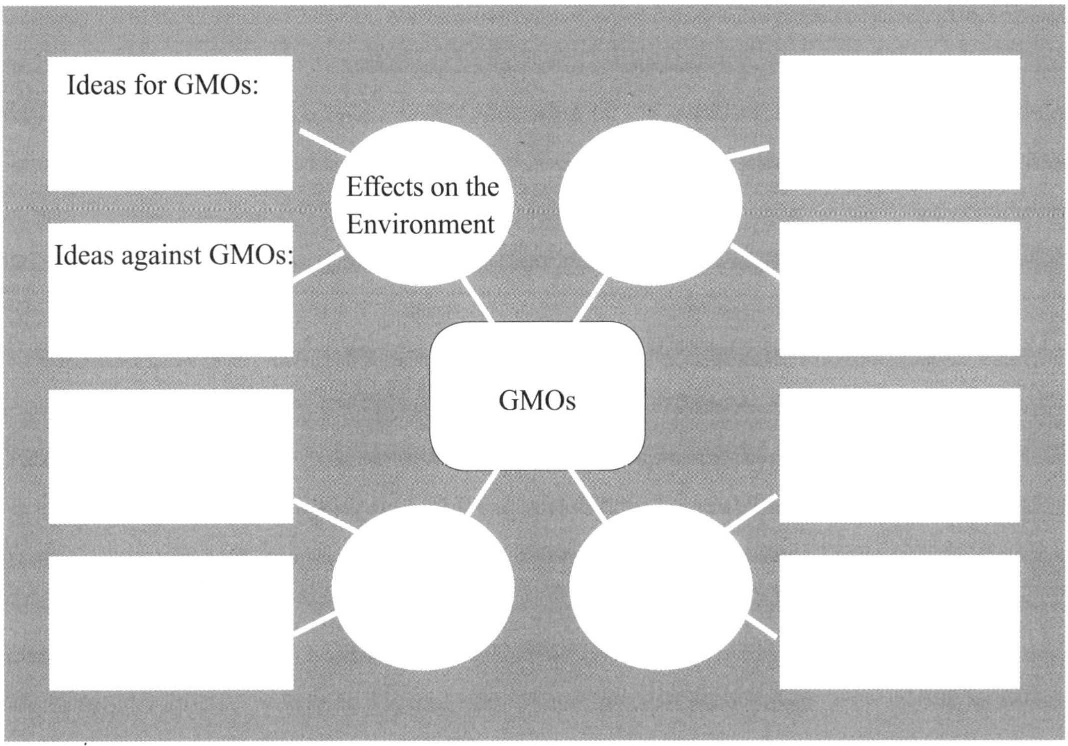

Text B

Analyze the information in each part of the passage, and then write a summary of each part.

Introduction:

Malaria's Deadly Toll:

____

____

How it All Begins—the Malaria Parasite Life Cycle:

____

____

Cucumbers that Eat themselves Alive and Malaria:

____

____

Cucumbers that Cure:

____

____

Part Cucumber and Part Fly:

____

____

Still No Solution (of should we say "dressing"):

____

____

Ⅱ. Paraphrasing

Interpret the following sentences in your own words.

1. The dizzying range of GMO categories is enough to boggle the mind. CRISPR, a novel genome editing tool, has allowed geneticists to breed GMO pigs that glow in the dark by inserting jellyfish bioluminescence genetic code into pig DNA.
2. Specifically, it's possible that modified organisms could be inbred with natural organisms, leading to the possible extinction of the original organism.
3. The argument over the development and marketing of GMO foods has become a political hot potato in recent years.
4. Some varieties of sea cucumbers are said to have excellent healing properties so it may be no surprise that this creature has found medicinal use, both formally and in alternative therapies. Extracts have found their way into oils, creams and cosmetics.
5. Furthermore, the genetically modified version of the mosquito would have to become the predominant species which is very difficult to achieve.

Ⅲ. Brief Answering

Answer the following questions based on information given in the two passages.

1. According to Text A, what are the potential risks associated with GM foods?
2. What are the reasons for people to welcome GMOs?
3. What are the reasons for people to be against GMOs?
4. According to Text B, what are the features of sea cucumbers which help in beating malaria?
5. How is genetic engineering used in fighting against malaria? And what are the problems in its application?

Ⅳ. Evaluation and Critical Thinking

Based on your analysis of the two passages, discuss in groups and present your ideas on the following items.

1. How would you comment on the organization of Text A?
2. In your opinion, argument given by which side in Text A is more convincing, Why? Assess your answer to the first question, and consider if you come to the conclusion in a scientific way.
3. What are your views on GMO labeling?
4. Make a flowchart on the life cycle of the malaria parasite, and figure out the point at which genetic engineering comes into play.
5. How would you evaluate genetic engineering based on what you have learned from the two passages?

Ⅴ. True/False Checking

Based on your understanding of the passages, decide whether each statement is true or false.

Text A

1. The principal application of techniques concerning GMOs is in agriculture.
2. GMOs are not safe to human health.
3. GM foods and foods made from crop plants are the same in terms of risks posed to human, according to the AAAS.
4. The fear for GM foods is more emotional than factual.
5. Currently there have not been federal requirements on labeling GMOs yet.

Text B

6. All mosquitoes are capable of transmitting parasite plasmodia.
7. Damaging the liver and blood cells can lead to flu-like symptoms.
8. A gene which determines the production of lectin and a gene which makes a substance that can be sent to the stomach are introduced to create the genetically modified mosquitoes.
9. The method of using GM mosquitoes is gaining momentum in the battle against malaria.
10. Commercial factors should be taken into consideration in the development of drug against malaria.

Ⅵ. Translation

Translate the following sentences into Chinese.

1. A genetically modified organism, or GMO, is an organism that has had its DNA altered or modified in some way through genetic engineering. (Para. 1, Text A)
2. GMO crops that are modified with the Bt gene have a proven resistance to insect pests, thus reducing the need for wide-scale spraying of synthetic pesticides. (Para. 9, Text A)
3. According to Christy Spees, a program manager with As You Sow, GMO foods are dangerous "because the modifications are centered around resistance to toxic substances, such as pesticides and certain fertilizers. When dangerous chemicals are applied, plants use them to grow, and the food itself can be detrimental to our health" .(Para. 14, Text A)
4. The disease causes an average loss of 1.3% annual economic growth in countries with a high incidence of infection. Furthermore, malaria has lifelong effects through increased poverty, impaired learning and decreased attendance in schools and the workplace. (Para. 5, Text B)
5. The concept of genetically modifying mosquitoes to disrupt parasite development prior to human infection is a novel approach to control the disease that is gaining recognition due to limited success elsewhere. (Para. 18, Text B)

Additional Reading

Scientists Can't Agree on Whether Genetically Modified-Mosquito Experiment Went Horribly Wrong

1. Biotech company released millions of genetically modified mosquitoes into Jacobina in Brazil.

2. From 2013 to 2015, an English biotech company released millions of genetically modified mosquitoes into neighborhoods in Jacobina, Brazil, in an effort to reduce the number of native disease-carrying mosquitoes. But unexpectedly, some of the gene-edited mosquitoes passed on their genes to the native insects, fueling concerns that they created a more robust hybrid species, according to new findings.
3. Considered the world's deadliest animal, mosquitoes spread a plethora of diseases, including Zika virus, dengue fever, yellow fever and West Nile virus.
4. To try to rid the world of some of these disease transmitters, a biotech company called Oxitec released around 450,000 genetically modified male Aedes aegyptimosquitoes into Jacobina each week for 27 months. These mosquitoes were altered such that they carried a "lethal gene" .
5. Once released, these ticking bombs were supposed to flit along and mate with females (the sex that bites humans) and then die, but not before they passed their lethal genes to similarly doomed offspring. In the lab, scientists had found that about 3% of the females that mated with the genetically modified males would produce offspring. But even the small number of offspring that survived were weak and unable to produce offspring of their own.
6. But now, a group of researchers not involved with Oxitec is raising questions as to whether this method went as planned. This method has successfully reduced native mosquito populations in Brazil by up to 85%, the researchers wrote.
7. They took genetic samples of the native population of mosquitoes in Brazil six, 12 and 27 to 30 months after the company released the genetically modified mosquitoes.
8. They found that some of the genes from the genetically modified mosquitoes had transferred to the native population. In other words, some of the offspring had survived and were strong enough to reproduce. This new population is a hybrid of Brazilian mosquitoes and the genetically modified mosquitoes that were created from strains in Cuba and Mexico, according to the study, which was published Sept. 10 in the journal Scientific Reports.
9. "The claim was that genes from the release strain would not get into the general population because offspring would die," senior author Jeffrey Powell, a professor of ecology and evolutionary biology at Yale University, said in a statement. "That obviously was not what happened."
10. However, there is no known health risk to humans that might come from these hybrids, he said. "But it is the unanticipated outcome that is concerning," Powell added.
11. In fact, the genes that were passed on weren't the tweaked genes that were designed to kill and tag the mosquitoes but rather genes from the strains in Cuba and Mexico, according to

Science magazine. The researchers also noted that this mixing of genes might have led to a "more robust population", perhaps one that would be better able to resist insecticides or transmit diseases, Science magazine reported.

12. That suggestion has prompted a backlash from the company. "We're not surprised by the results, but what we are surprised by are the speculations that the authors have made," Nathan Rose, head of scientific and regulatory affairs at Oxitec, told Science magazine.

13. The company has requested that the journal review these "misleading and speculative statements"; indeed, the paper now includes an editor's note saying the journal is considering the criticisms.

Unit Nine Space Exploration

Critical Reading

Text A

The Alignment of Planets

1. “Beware the Ides of March,” the crone intoned to the Roman dictator in 44 B.C. .But Caesar, secure in his divinity and power, ignored her and shortly thereafter was sent from this Earth by some of his closest “friends”. Thc position of hcavcnly objects played a role in the assassination because, by most accounts, it was an astrologer who foretold his demise. Emboldened by her prediction, Caesar’s assassins turned it into a self-fulfilling prophecy.
2. “There was a similar case about 140 years after Caesar met his end,” says Florian Himmler, a researcher in ancient history at the University of Regensburg in Bavaria, Germany. On September 18, 96 A.D., the Roman emperor Titus Flavius Domitianus was also sent packing by assassins—some were his closest friends and courtiers. His assassins chose the date and hour of his departure based upon the position of the planets … including Mars, which was positioned to make his ‘divine protection’ weakest.
3. Centuries ago monitoring the stars and planets was a popular way to plan daily events. Some say it still is! But the scientific method has shown that astrology holds little, if any, predictive power. As a result the belief in astrology is now far less universal than it was in Titus’ day.
4. Nevertheless there are certain endeavours that are absolutely dependent upon the positions of the planets. In fact, some of our civilisations most advanced organisations, like NASA and other space agencies around the world, sometimes do nothing without first consulting the stars!
5. In this case, however, it’s not for luck. NASA’s mission planners carefully check the heavens to assure that their targets—usually planets, comets or asteroids—are in the right place to make journeys there as short and inexpensive as possible.
6. Such checks are rarely done in science fiction. When Star Trek’s Captain Kirk wants to go

someplace he never waits for a propitious alignment—he just points the Enterprise in the right direction and cries "Warp Speed, Sulu!" Or in Star Wars, when Han Solo wants to travel to the Alderaan District, he simply pushes a few buttons and off he goes.

7. Unlike the mighty vessels of Kirk and Solo, however, our present-day space ships harbour limited power. Even the awesome Saturn V rocket, which carried 45,000 kg to lunar orbit during the Apollo program, didn't completely escape the pull of Earth's gravity. (Remember, the Moon is trapped by our planet's gravitational field and that's as far as the Saturn V went.) Nowadays the space shuttle can haul about 25,000 kg into low Earth orbit. Without extra propulsion built in, however, those payloads are still tightly bound to Earth's gravitational field.

8. Of course, some real-life spacecraft can reach escape velocity and travel to other worlds. Delta 2 rockets—often used to send missions to Mars—can loft about 700 kilograms free of Earth's gravity. But we can't send those 700 kg anywhere we want, for two reasons. First, such payloads remain bound to the Sun's gravitational field. Even after escaping Earth, they are still trapped within the solar system! Second, once the rocket engine exhausts its fuel, which happens quickly for chemical rockets, the payload can do little but coast in the direction it was slung.

9. Interplanetary coasting can take a long time. The recently-launched 2001 Mars Odyssey, for instance, will reach the Red Planet fully six months after it left Earth. During that interval Mars will have moved one-quarter of the way around its orbit. Clearly, it's vital that we understand not only where the target is at launch, but also where it will be when the spacecraft arrives. Present-day astronomers and mission planners find themselves calculating planetary motions and alignments much as their ancient ancestors did!

10. NASA has been considering a human mission to Mars for years. Larry Kos, a mission planner at NASA's Marshall Space Flight Centre, notes that timing is everything. "The best time to launch a mission to Mars," he says, "is usually a few months before Earth and Mars are closest together—a time astronomers call 'opposition'. When Mars missions take off, they head toward an apparently empty point in space. The planet isn't there yet, but it will be when the spacecraft arrives." Of course, if humans go to Mars they will need to come back, too. "For a return trip we would wait 26 months for a similar Earth-Mars alignment and once again launch a few months before opposition. That geometry would minimise the return propulsion needed."

11. While Earth and Mars approach each other every 26 months, their minimum separation varies over a 15 year cycle due to the elliptical nature of each planet's orbit. Indeed, it can vary by

almost a factor of two. Choosing the right year to launch will have a significant impact on the propulsion power required to fling a payload from Earth to Mars, and back again.

12. The next best times to go to Mars will come in 2003, 2018, and 2020—years when Earth and Mars will be unusually close together. Humans might finally visit the Red Planet in 2018 or 2020, but alas, they won't travel there aboard vessels like the USS Enterprise or the Millennium Falcon. Our first Martian explorers will probably blast off on chemical rockets after intensive calculations of capability, aim points, and timing. In that regard, human exploration of Mars will begin as have so many other adventures in history … only when the planets are properly aligned.

(918 words)

Glossary:

alignment *n.* apparent meeting or passing of two or more celestial bodies in the same degree of the zodiac（天文学）行星等的会合

astrology *n.* a pseudoscience claiming divination by the positions of the planets and sun and moon 占星术

coast *v.* move effortlessly 滑翔

consult *v.* seek information from 咨询

elliptical *adj.* rounded like an egg 椭圆的

intone *v.* speak carefully in a particular tone 以特殊的语调说

payload *n.* goods carried by a large vehicle 负载

propulsion *n.* force causing to move forward 推动力

sling *v.* hurl 投掷

Text B

Spaceships of the Future (Abridged)

1. It may be the oldest cliché in town, but in the not too distant future science fiction will turn into science fact. The fantastic spaceships of sci-fi comic books and novels will no longer be a figment of our creative imagination; they may be the real vision of our future.

2. Engineers and designers are already designing craft capable of propelling us beyond Earth's orbit, the Moon and the planets. They're designing interstellar spaceships capable of travel across the vast emptiness of deep space to distant stars and new planets in our unending quest

to conquer and discover. Our Universe contains over a billion galaxies; star cities each with a hundred billion inhabitants. Around these stars must exist planets and perhaps life. The temptation to explore these new realms is too great.

3. First things first—we'll have to build either a giant orbiting launch platform, far bigger than the International Space Station (ISS), or a permanently manned lunar base to provide a springboard for the stars. Some planners feel we should limit ourselves to robotic probes, but others are firmly committed to sending humans. "There's a debate right now about how to explore space," says astronaut Bill Shepherd, destined to be the first live-aboard Commander of the ISS. "Humans or machines—I think they're complementary."

The Human Problem

4. Space is the most hostile environment we will ever explore. Even a single five-hour spacewalk requires months of training, and a vast technical backup to keep it safe. The astronauts and cosmonauts who live aboard the ISS will be there for only a few weeks or months; if we want to travel into deep space it could take years. First we'll have to find out just how long the human body can survive in a weightless environment. In zero gravity, four pints of body fluid rush from the legs to the head where it stays for the duration of the mission. Astronauts often feel as if they have a permanent cold, and disorientation can become a major problem. In space there's no physical sensation to let you know when you're upside down and astronauts have to rely on visual clues from their surroundings. A few hours after reaching orbit, one in three of all astronauts will experience space sickness—a feeling rather like carsickness. And weightless conditions lead to calcium being leached from the bones, and problems with the astronauts' immune systems.

5. Trillions of rocky fragments—meteoroids—roam our Solar System at speeds of up to 150,000 miles an hour. A meteoroid no bigger than a grain of salt could pierce a spaceship window. Protection from the extreme hazards of space is going to need some clever technology. Space is also full of lethal radiation—X-rays, gamma rays and the high-speed particles called cosmic rays.

6. Down here on Earth we are protected by the atmosphere and by our planet's magnetic field, but in space long haul astronauts suffer gradual but irreversible radiation sickness unless they are carefully shielded. Commander Shepherd is confident the ISS will help us crack the problems. "The ISS is going to answer a number of questions about long range exploration in space. A lot of things are going to be pioneered on the space station for future exploration."

It's Only Rocket Science

7. Scientists are already experimenting with propulsion systems that may travel much faster than today's conventional chemical rockets. Franklin Chang's plasma rocket may be the answer. "In a plasma rocket you're continually accelerating," he explains.
8. A trip to Mars could be cut to 90 days, claims Chang. His rocket harnesses a nuclear process to produce a hot gas plasma. The plasma is magnetically held in a rocket the shape of a bottle and then expelled at very high velocity to provide propulsion. The plasma has to be heated to millions of degrees. Chang believes his system will be too good just to reach Mars. "I think it will quickly be developed for interplanetary travel within our Solar System." The plasma rocket is now under development at NASA's Houston laboratories.
9. Another new method of propulsion is already flying through our Solar System. Pushed only by an electronically driven 'ion engine', Deep Space One is already over 100 million miles from Earth. It works by ionising xenon gas and expelling it with the aid of electric fields, so providing a gentle but constant thrust. The ion engine provides a force about the same as a single sheet of paper exerts on your hand—far too weak to lift a spacecraft from the surface of a planet—but the continuous acceleration has already pushed Deep Space One to a speed ten times higher than any of the manned rockets we use today.

Interstellar Travel

10. To leave the Solar System and carry humans to the stars we will have to find a way of travelling near to the speed of light. Even then a journey could take hundreds or thousands of years. Travelling at 1/10 the speed of light it would take over forty years to reach the nearest star, Alpha Centauri.
11. One giant source of free energy is our Sun. Bob Forward has designed the solar sail, a craft that doesn't have to carry its own fuel supply. It's driven by the power of the Sun's rays, and it will be the fastest machine ever built. "The sunlight bounces off the aluminium sails and in the process gives it a tiny push," explains Forward. Like the ion probe it will accelerate and accelerate. And it's not a total dream. NASA is already experimenting with deploying large sails in Earth-orbit. Propelled by light, solar sails will travel thousands of times faster than Apollo or the Shuttle.

Asleep or Awake?

12. Even with the perfect spaceship it isn't going to be easy. In his classic sci-fi novel 2001, Arthur C. Clarke used the concept of suspended animation as a way for humans to cope with long space flights. He imagined that we would be able to put the human body into hibernation —suspended animation—to escape the boredom of long interstellar missions.

13. An even more drastic measure might be to freeze the astronauts. We already use cryogenic techniques to preserve dead bodies and store human embryos. Freezing living adults may not be so far away, but perhaps we won't have to do that. Perhaps we should use our existing technology and send frozen embryos across to the far corners of the cosmos.

14. It could certainly save on space. Then hundreds of years from now, billions and billions of miles away, the embryos will be thawed and their hearts will start beating. These space-farers of the future will not grow inside a mother's body but will be incubated in a machine. They will be brought up by robot. It may seem strange and radical but one day it might just happen.

15. "Who's to say in 25 years what we'll be doing in space? I think all estimates may be wildly short of the mark," muses Commander Shepherd.

(1,280 words)

Glossary:

acceleration *n.* increasing the speed 加速

complementary *adj.* acting as or providing a something that completes the whole 互补的

cryogenic *adj.* of or relating to very low temperatures 低温的

deploy *v.* distribute systematically or strategically 部署

disorientation *n.* uncertainty as to direction 定位障碍

figment *n.* a contrived or fantastic idea 虚构之物

harness *v.* control and direct 控制

incubate *v.* grow under conditions that promote development 培养

plasma *n.* a fourth state of matter distinct from solid or liquid or gas and present in stars and fusion reactors 等离子体

thaw *v.* cause to become soft or liquid 使融解

Exercises

Ⅰ. Outline and Summary

Read the passages and finish the outline and summary exercises.

Text A

Analyze the main idea of each part and find out how the writer organizes this passage. Figure out the logical connection between each part, and work out how coherence and cohesion are achieved.

Para. 1 ~ 3	Ancient anecdotes concerning observation of stars
Para. 4 ~ 6	(1) _________ observe the stars for none of the same reason as the ancestors, nor as in science fictions
Para. 7 ~ 9	The reason why NASA checks the (2) _________of planets
Para. 10 ~ 12	The best time to (3) _________

Text B

Analyze the information given in each section of the passage, and write a summary of each section in one sentence.

The human problem:

____

____

It's only rocket science:

____

____

Interstellar travel:

____

____

Asleep or awake?:

____

____

Ⅱ. Paraphrasing

Interpret the following sentences in your own words.

1. "Beware the Ides of March," the crone intoned to the Roman dictator in 44 B.C.. But Caesar, secure in his divinity and power, ignored her and shortly thereafter was sent from this Earth by some of his closest "friends".
2. Without extra propulsion built in, however, those payloads are still tightly bound to Earth's gravitational field.
3. Present-day astronomers and mission planners find themselves calculating planetary motions and alignments much as their ancient ancestors did!
4. Space is the most hostile environment we will ever explore. Even a single five-hour spacewalk requires months of training, and a vast technical backup to keep it safe.
5. "Who's to say in 25 years what we'll be doing in space? I think all estimates may be wildly short of the mark," muses Commander Shepherd.

Ⅲ. Brief Answering

Answer the following questions based on information given in the two passages.

1. According to Text A, what is the purpose of NASA's observation of stars?
2. What do the ancient ancestors and NASA have in common in terms of the observation of stars?
3. What is the significance of choosing a right year to launch a mission to Mars?
4. According to Text B, what needs to be done first for taking a space travel?
5. What are the sources of energy for taking interstellar travel?

Ⅳ. Evaluation and Critical Thinking

Based on your analysis of the two passages, discuss in groups and present your ideas on the following items.

1. In Text A, what functions do the anecdotes of Caesar and Domitianus play? And why are Star Trek and Star Wars mentioned in the development of the passage?
2. Based on information given in Text A, why can't we send spaceships anywhere we want?
3. How would you comment on the writing of the conclusion of Text A?
4. What do you think are other aspects concerning spaceships in the future which the author has not covered in Text B?
5. After studying Text A and Text B, how would you comment on man's probe into space from

the past to the future?

Ⅴ. True/False Checking

Based on your understanding of the passages, decide whether each statement is true or false.

Text A

1. In ancient times, stars played a more astrological role than an astronomical one.
2. There are economic reasons for NASA to observe stars.
3. Spacecrafts have to reach escape velocity in order to break away from the Earth's gravitational field.
4. "Opposition" is the optimal time to launch a mission to Mars.
5. Alignment of planets matters in launching a mission to Mars.

Text B

6. There are about a hundred billion living beings in galaxies in our universe.
7. In the outer space, any tiny particle the size of a grain of salt will do harm to a spacecraft.
8. Deep Space One is at a speed higher than any of the manned rockets.
9. The solar sail is a craft driven by sun light.
10. Astronauts will be frozen to escape the boredom of long interstellar missions.

Ⅵ. Translation

Translate the following sentences into Chinese.

1. Nevertheless there are certain endeavours that are absolutely dependent upon the positions of the planets. In fact, some of our civilisations most advanced organisations, like NASA and other space agencies around the world, sometimes do nothing without first consulting the stars! (Para. 4, Text A)
2. Unlike the mighty vessels of Kirk and Solo, however, our present-day space ships harbour limited power. Even the awesome Saturn V rocket, which carried 45,000 kg to lunar orbit during the Apollo program, didn't completely escape the pull of Earth's gravity. (Para. 7, Text A)
3. Our first Martian explorers will probably blast off on chemical rockets after intensive calculations of capability, aim points, and timing. In that regard, human exploration of Mars will begin as have so many other adventures in history … only when the planets are properly

aligned. (Para. 12, Text A)

4. They're designing interstellar spaceships capable of travel across the vast emptiness of deep space to distant stars and new planets in our unending quest to conquer and discover. (Para. 2, Text B)
5. NASA is already experimenting with deploying large sails in Earth-orbit. Propelled by light, solar sails will travel thousands of times faster than Apollo or the Shuttle. (Para. 11, Text B)

Additional Reading

Chinese Moon Rover Peers beneath Surface of Mysterious Lunar Far Side

1. The Chang'e 4 spacecraft touched down on the floor of the 115-mile-wide (186 kilometers) Von Kármán Crater on Jan. 2, 2019, becoming the first probe ever to ace a soft landing on the moon's mysterious far side, which forever points away from Earth.
2. A rover called Yutu 2 ("Jade Rabbit 2") rolled off the stationary Chang'e 4 lander just hours after touchdown. These two solar-powered craft have now been taking the measure of their exotic surroundings for more than a year with a variety of science gear, giving us unprecedented views of the lunar far side's surface.
3. Those views now extend to the subsurface, thanks to the first published results from Yutu 2's ground-penetrating radar instrument. In a paper released Wednesday (Feb. 26) in the journal Science Advances, Chang'e 4 scientists revealed the structure of the gray dirt beneath the rover's wheels, as gleaned from radar data gathered during Yutu 2's first two lunar days of operation. (Each lunar day is about two Earth weeks long. Yutu 2 and the Chang'e 4 lander hibernate during the brutally cold lunar nights, which also last two weeks apiece.)
4. The researchers discerned three distinct layers beneath Yutu 2's section of the Von Kármán floor. The top layer, which extends about 39 feet (12 meters) down, consists of relatively uniform lunar regolith (soil), with a few large rocks mixed in here and there. The second layer, which goes from about 39 feet to 79 feet (24 m) deep, harbors coarser-grained materials and greater numbers of embedded rocks. The third stratum, which goes down to at least 130 feet (40 m)—the radar couldn't reliably penetrate any deeper—features alternating bands of coarse and fine-grained material, as well as embedded rocks.
5. The data indicate "that the subsurface internal structure at the landing site is essentially made by low-loss, highly porous granular materials embedding boulders of different sizes", the

researchers, led by Chunlai Li of the National Astronomical Observatories at the Chinese Academy of Sciences and the University of Chinese Academy of Sciences in Beijing, wrote in the new paper.

6. And this structure tells us about the region's violent history, the scientists added: "Given such a strong geological constraint, the most plausible interpretation is that the sequence is made of a layer of regolith overlaying a sequence of ejecta deposits from various craters, which progressively accumulated after the emplacement of the mare basalts on the floor of Von Kármán Crater."
7. Those dark mare basalts are evidence of ancient volcanic activity, which flooded Von Kármán with molten rock about 3.6 billion years ago. ("Mare" is Italian for "sea". Early astronomers thought that these big patches on the moon were bodies of water.)
8. Yutu 2's newly published observations also show that ground-penetrating radar can be a valuable tool for other lunar surface craft going forward, especially on the little-known far side, the researchers said.
9. Extensive use of this tool "could greatly improve our understanding of the history of lunar impact and volcanism and could shed new light on the comprehension of the geological evolution of the moon's far side", the researchers wrote.
10. The Chang'e 4 lander was originally designed to operate for 12 months and Yutu 2 for just three months. Both craft have therefore already exceeded their warranties, and both are still going strong; the duo recently woke up to begin their 15th lunar day of science work.
11. Chang'e 4 is part of China's ambitious Chang'e program of robotic lunar exploration, which is named after a Chinese moon goddess. The next mission up, the Chang'e 5 sample-return effort, is scheduled to launch later this year.

Unit Ten Science Mystery

Critical Reading

Text A

The Mystery of the Bermuda Triangle

1. Miami, Puerto Rico and Bermuda are prime holiday destinations boasting sun, beaches and coral seas. But between these idyllic settings, there is a dark side: countless ships and planes have mysteriously gone missing in the one and a half million square miles of ocean separating them. About 60 years ago, the area was claiming about five planes every day and was nicknamed the Bermuda Triangle by a magazine in 1964. Today, about that many planes disappear in the region each year and there are a number of theories explaining what could be happening.
2. Twins George and David Rothschild are among the first passengers to have experienced bizarre effects in the Bermuda Triangle. In 1952, when they were 19 years old, the two naval men had to make an emergency trip home on a navy light aircraft, north over the Florida Keys, to attend their father's funeral. "We had been flying for probably 20 or 30 minutes when all of a sudden the pilot yelled out that the instruments were dead and he became very frantic," says George Rothschild. He had lost his bearings, and not only did he not know where he was, he also had no idea how much gas was left in the fuel tanks. After what seemed like hours, they landed safely in Norfolk, on the Florida coast.
3. Some speculate that it had nothing to do with the location, but rather the instruments that were available at the time. Pilot Robert Grant says that back in the 1940s, navigating a plane involved a lot of guesswork since they relied completely on a magnetic compass to guide them. Dead reckoning was used, which means that pilots would trust their compass and then estimate how the wind would influence their planned flight path to remain on track. No matter what your mind tells you, you must stay on that course, says Grant. If you don't, and you start

turning to wherever you think you should be going, then you're toast.

Wild Weather

4. The landscape of the island of Bermuda is quite unique: it is a remote coral reef precariously perched on a massive extinct volcano. Fisherman Sloan Wakefield, who knows the waters of the Bermuda Triangle very well, thinks that the weather could be responsible for some of the disappearances. Because the island is a dot in the Atlantic Ocean, it gets weather from everywhere and it can change in a heartbeat. One minute, you can be looking at good weather, and the next moment you've got a low front coming through, he says. He has already seen 15 to 20 foot (4.6 m to 6 m) waves on the sea.
5. Hurricanes are common in the Bermuda Triangle area. In the Atlantic Ocean, they typically originate off the African coast and thrive off the moisture of the warm, tropical waters. Hurricane records from the past 100 years have shown that they often head west for the United States but swerve into the waters of the Bermuda Triangle at the last minute. Jim Lushine, a meteorologist at thc National Hurricane Centre in Miami, Florida studies the weather in the Bermuda Triangle and says that there are more hurricanes in that particular area than in any other in the Atlantic basin.
6. But thunderstorms in the area can be just as dangerous. In 1986, a historic ship called the Pride of Baltimore vanished from radar screens while it was in the Bermuda Triangle, making a trip from the Caribbean to Baltimore. About four and a half days later, the wreckagc and eight survivors were found and they revealed that the ship had been hit by a microburst: 80 mile per hour winds emanating from a freak thunderstorm. It happened so quickly that the crew didn't have time to make a distress call. The ship was sunk in the downburst, unfortunately with a great loss of life, says Lushine. Similar downbursts are probably responsible for some of the sunken ships in the Bermuda Triangle.
7. Even more unpredictable than thunderstorms are waterspouts. These can be caused by tornadoes that move out to sea or rotating columns of air that drop from thunderstorms, creating a vortex of spray. When the moisture condenses, it forms a twisting column that connects the sea to the clouds. Jim Edds, an amateur fisherman who chases and films waterspouts for fun, says that if you are out at night and a tornado-like waterspout develops—the really big, strong ones with high velocity—it can flip your vessel over.

Bubbling Methane

8. Seismic activity at the bottom of the ocean can also be an explanation for disappearing ships. Scientists have discovered that huge bubbles of methane gas can violently erupt without warning from the ocean floor and at least one oil rig is thought to have sunk because of this phenomenon. Ralph Richardson, the director of the Bermuda Underwater Exploration Institute, claims that a large pocket of gas could surround a ship, causing it to lose buoyancy and disappear without warning.
9. At the U.S. Navy's research centre in California, Bruce Denardo, an expert in fluid dynamics, has proved that bubbles from methane gas eruptions could be responsible for vanishing ships in the open ocean. Water pressure causes objects to float, and the deeper the water, the greater the pressure exerted to keep the object floating at the surface. If bubbles from methane are introduced, they lower the density of the water. They take up space, but the volume of water stays the same, causing the buoyant force to decrease. In an experiment with a ball in water, Denardo can demonstrate that the ball sinks deeper and deeper down in water as the amount of bubbles increases, until it reaches a critical point where it sinks completely. If a ship were to take on enough water, it would sink to the bottom and stay there, says Denardo.

A Mysterious Time Warp?

10. Others have more far out explanations for the Bermuda Triangle disappearances. Property developer Bruce Gernon claims that on December 4th 1970, when he flew from the island of Andross in the Bahamas to Florida, he experienced a distortion in space time. He had made the same trip on many occasions, but he claims that his journey that day was much faster than usual. "I noticed a huge U-shaped opening in the clouds, but as I approached it, the top of the opening closed and it became a horizontal tunnel that appeared to be 10 to 15 miles long," he says. "When the aircraft entered the tunnel, some lines, which I call time lines, appeared which were rotating counter-clockwise. It was difficult to keep it level and concentrate on the other end of the tunnel which was aiming directly for Miami. "
11. Gernon claims that when he came out of the tunnel, it closed fast behind him and he was surrounded by a strange fog. His instruments had stopped working and Air Traffic Control had no radar trace of his plane until they realized that it was actually over Miami beach. Given the time they had been flying, they should still have been about 45 minutes away from Miami. After researching what could have happened, Genon is now writing a book about his experience. "I have come to the conclusion that we experienced a space time warp of a

hundred miles in thirty minutes," he says.

12. Is this scientifically feasible? About 80 years ago, Einstein proposed his general theory of relativity which claimed that huge spinning objects could distort space and time in their surroundings. Although NASA researchers have now found signs that black holes and neutron stars appear to warp space time, it is still a far cry from concepts introduced by science fiction like wormholes, or tunnels in space time that provide travellers with an express route between different dimensions and great distances.

13. Explanations for the vanishings in the Bermuda Triangle are all still theories. But especially for people who have witnessed bizarre events in this area, there is a strong desire to find some answers. One author, Gian Quasar, has been investigating every single plane and ship disappearance in the Bermuda Triangle and has listed every case on a massive internet database at http://www.bermuda-triangle.org/. With initiatives like this and further research, perhaps the mystery will come to a conclusion.

(1,384 words)

Glossary:

buoyancy *n.* the tendency to float in water or other liquid 浮力

dead Reckoning *n.* navigation without the aid of celestial observations 推算航行法

emanate *v.* proceed or issue forth, as from a source 发出

meteorologist *n.* a specialist who studies processes in the carth's atmosphere that cause weather conditions 气象学家

precariously *adv.* in a way that is likely to fall, be damaged, fail, etc 不稳固地

swerve *v.* change direction abruptly 突然转向

seismic *adj.* subject to or caused by an earthquake or earth vibration 地震的

vortex *n.* the shape of something rotating rapidly 漩涡

warp *n.* a twist or aberration, especially a perverse or abnormal way of judging or acting 弯曲

Text B

The Biggest Unsolved Mysteries in Physics (Abridged)

Introduction

1. In 1900, the British physicist Lord Kelvin is said to have pronounced: "There is nothing new to be discovered in physics now. All that remains is more and more precise measurement." Within

three decades, quantum mechanics and Einstein's theory of relativity had revolutionized the field. Today, no physicist would dare assert that our physical knowledge of the universe is near completion. To the contrary, each new discovery seems to unlock a Pandora's box of even bigger, even deeper physics questions. These are our picks for the most profound open questions of all.

2. Inside you'll learn about parallel universes, why time seems to move in one direction only, and why we don't understand chaos.

3. Editor's Note: This list was originally published in 2012. It was updated on Feb. 27, 2017, to include newer information and recent studies.

What Is Dark Energy?

4. No matter how astrophysicists crunch the numbers, the universe simply doesn't add up. Even though gravity is pulling inward on space-time—the "fabric" of the cosmos—it keeps expanding outward faster and faster. To account for this, astrophysicists have proposed an invisible agent that counteracts gravity by pushing space-time apart. They call it dark energy. In the most widely accepted model of dark energy, it is a "cosmological constant": an inherent property of space itself, which has "negative pressure" driving space apart. As space expands, more space is created, and with it, more dark energy. Based on the observed rate of expansion, scientists know that the sum of all the dark energy must make up more than 70 percent of the total contents of the universe. But no one knows how to look for it. The best researchers have been able to do in recent years is narrow in a bit on where dark energy might be hiding, which was the topic of a study released in August 2015.

What Is Dark Matter?

5. Evidently, about 84 percent of the matter in the universe does not absorb or emit light. "Dark matter," as it is called, cannot be seen directly, and it hasn't yet been detected by indirect means, either. Instead, dark matter's existence and properties are inferred from its gravitational effects on visible matter, radiation and the structure of the universe. This shadowy substance is thought to pervade the outskirts of galaxies, and may be composed of "weakly interacting massive particles", or WIMPs. Worldwide, there are several detectors on the lookout for WIMPs, but so far, not one has been found. One recent study suggests dark mater might form long, fine-grained streams throughout the universe, and that such streams might radiate out from Earth like hairs.

Why Is There An Arrow of Time?

6. Time moves forward because a property of the universe called "entropy", roughly defined as the level of disorder, only increases, and so there is no way to reverse a rise in entropy after it has occurred. The fact that entropy increases is a matter of logic: There are more disordered arrangements of particles than there are ordered arrangements, and so as things change, they tend to fall into disarray. But the underlying question here is, why was entropy so low in the past? Put differently, why was the universe so ordered at its beginning, when a huge amount of energy was crammed together in a small amount of space?

Are There Parallel Universes?

7. Astrophysical data suggests space-time might be "flat", rather than curved, and thus that it goes on forever. If so, then the region we can see (which we think of as "the universe") is just one patch in an infinitely large "quilted multiverse". At the same time, the laws of quantum mechanics dictate that there are only a finite number of possible particle configurations within each cosmic patch (10^10^122 distinct possibilities). So, with an infinite number of cosmic patches, the particle arrangements within them are forced to repeat—infinitely many times over. This means there are infinitely many parallel universes: cosmic patches exactly the same as ours (containing someone exactly like you), as well as patches that differ by just one particle's position, patches that differ by two particles' positions, and so on down to patches that are totally different from ours.
8. Is there something wrong with that logic, or is its bizarre outcome true? And if it is true, how might we ever detect the presence of parallel universes? Check out this excellent perspective from 2015 that looks into what "infinite universes" would mean.

Why Is There More Matter than Antimatter?

9. The question of why there is so much more matter than its oppositely-charged and oppositely-spinning twin, antimatter, is actually a question of why anything exists at all. One assumes the universe would treat matter and antimatter symmetrically, and thus that, at the moment of the Big Bang, equal amounts of matter and antimatter should have been produced. But if that had happened, there would have been a total annihilation of both: Protons would have canceled with antiprotons, electrons with anti-electrons (positrons), neutrons with antineutrons, and so on, leaving behind a dull sea of photons in a matterless expanse. For some reason, there was excess matter that didn't get annihilated, and here we are. For this, there is no accepted

explanation. The most detailed test to date of the differences between matter and antimatter, announced in August 2015, confirm they are mirror images of each other, providing exactly zero new paths toward understanding the mystery of why matter is far more common.

What Happens inside A Black Hole?

10. What happens to an object's information if it gets sucked into a black hole? According to the current theories, if you were to drop a cube of iron into a black hole, there would be no way to retrieve any of that information. That's because a black hole's gravity is so strong that its escape velocity is faster than light—and light is the fastest thing there is. However, a branch of science called quantum mechanics says that quantum information can't be destroyed. "If you annihilate this information somehow, something goes haywire," said Robert McNees, an associate professor of physics at Loyola University Chicago.

11. Quantum information is a bit different from the information we store as 1 s and 0 s on a computer, or the stuff in our brains. That's because quantum theories don't provide exact information about, for instance, where an object will be, like calculating the trajectory of a baseball in mechanics. Instead, such theories reveal the most likely location or the most likely result of some action. As a consequence, all of the probabilities of various events should add up to 1, or 100 percent. (For instance, when you roll a six-sided die, the chances of a given face coming up is one-sixth, so the probabilities of all the faces add up to 1, and you can't be more than 100 percent certain something will happen.) Quantum theory is, therefore, called unitary. If you know how a system ends, you can calculate how it began.

12. To describe a black hole, all you need is mass, angular momentum (if it's spinning) and charge. Nothing comes out of a black hole except a slow trickle of thermal radiation called Hawking radiation. As far as anyone knows, there's no way to do that reverse calculation to figure out what the black hole actually gobbled up. The information is destroyed. However, quantum theory says that information can't be completely out of reach. Therein lies the "information paradox".

13. McNees said there has been a lot of work on the subject, notably by Stephen Hawking and Stephen Perry, who suggested in 2015 that, rather than being stored within the deep clutches of a black hole, the information remains on its boundary, called the event horizon. Many others have attempted to solve the paradox. Thus far, physicists can't agree on the explanation, and they're likely to disagree for some time.

(1,316 words)

Glossary:

annihilate *v.* destroy something completely so that nothing is left 湮灭

astrophysicist *n.* an astronomer who studies the physical properties of celestial bodies 天体物理学家

configuration *n.* an arrangement of parts or elements 结构，布局

cosmological *adj.* pertaining to the branch of astronomy dealing with the origin and history and structure and dynamics of the universe 宇宙论的

counteract *v.* act against and reduce the force or effect of 抵消

cram *v.* force into a small space 填塞

entropy *n.* a numerical measure of the uncertainty of an outcome 熵

trajectory *n.* the path followed by an object moving through space 轨道

paradox *n.* a statement that contradicts itself 悖论

Exercises

Ⅰ. Summary and Mindmap

Read the passages and finish the summary and mindmap exercises.

Text A

Analyze the passage and summary the information given in each part of it.

Introduction:

____

____

Wild Weather:

____

____

Bubbling Methane:

____

____

A Mysterious Time Warp:

____

____

Conclusion:

____

____

Text B

Analyze the information given in each part, categorize it and complete the following mindmap.

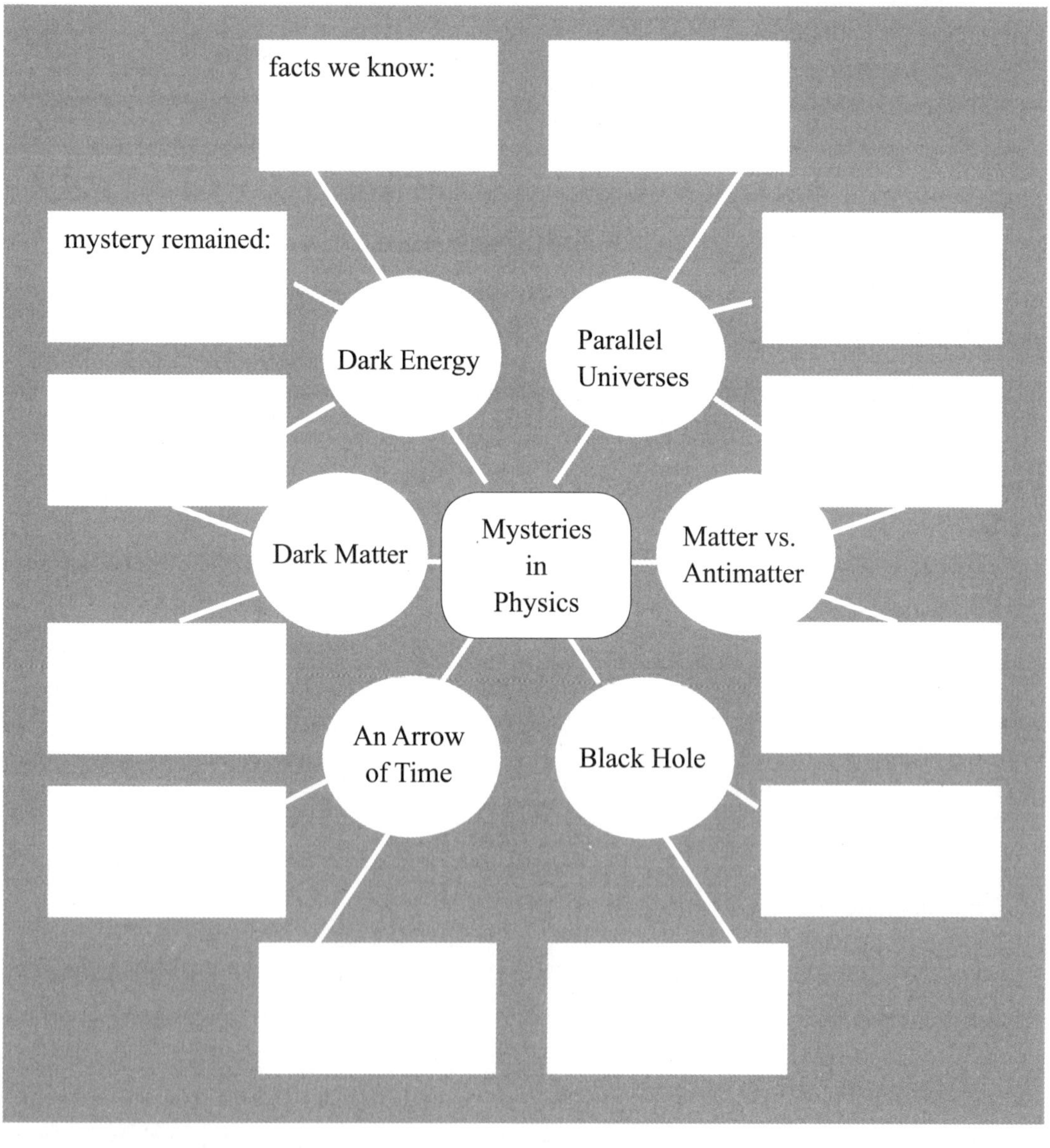

Ⅱ. Paraphrasing

Interpret the following sentences in your own words.

1. But between these idyllic settings, there is a dark side: countless ships and planes have mysteriously gone missing in the one and a half million square miles of ocean separating them.
2. In the Atlantic Ocean, they typically originate off the African coast and thrive off the moisture of the warm, tropical waters. Hurricane records from the past 100 years have shown that they often head west for the United States but swerve into the waters of the Bermuda Triangle at the last minute.
3. Others have more far out explanations for the Bermuda Triangle disappearances.
4. This shadowy substance is thought to pervade the outskirts of galaxies, and may be composed of "weakly interacting massive particles", or WIMPs.
5. According to the current theories, if you were to drop a cube of iron into a black hole, there would be no way to retrieve any of that information.

Ⅲ. Brief Answering

Answer the following questions based on information given in the two passages.

1. According to Text A, how does wild weather lead to vanishings in the Bermuda Triangle?
2. How does bubbling methane give rise to the sinking of vessels in the Bermuda Triangle?
3. What happened to Bruce Gernon on December 4th, 1970?
4. What is the logic of entropy increase?
5. What is the "information paradox"?

Ⅳ. Evaluation and Critical Thinking

Based on your analysis of the two passages, discuss in groups and present your ideas on the following items.

1. How would you comment on the way the author develops his description about the Bermuda Triangle in the introduction of Text A?
2. Make a judgment on the credibility of the three explanations for the mystery of the Bermuda Triangle, and then answer the questions: Which is the most convincing and which is the least? What is the basis of your judgment? Do you have a scientific attitude towards these explanations?
3. What do you think of the conclusion of Text A?

4. How is the beginning of Text B different from that of Text A? For what reasons do you think the two authors differ in opening up their essays?
5. After studying these texts, what tips would you offer to someone who is to write an essay explaining science mysteries to the public?

Ⅴ. True/False Checking

Based on your understanding of the passages, decide whether each statement is true or false.

Text A

1. In the incident, the pilot who flew the plane George and David were taking lost his instrument.
2. There are more hurricanes in the Bermuda Triangle because of the warm, tropical waters there.
3. Bubbling methane causes decrease in buoyancy which functions to sink a ship.
4. If Gernon were given 45 more minutes, he would have arrived in Miami on time.
5. Only people who have witnessed bizarre events in the Bermuda Triangle have strong desire to find the answer to the mystery.

Text B

6. New discoveries in physics today usually lead to questions that are even harder to explain.
7. Nothing else can rival dark energy in making up the universe.
8. Matter and antimatter are parallel in number.
9. The universe is getting more and more disordered.
10. Hawking radiation is the only thing that can escape from a black hole.

Ⅵ. Translation

Translate the following sentences into Chinese.

1. Miami, Puerto Rico and Bermuda are prime holiday destinations boasting sun, beaches and coral seas. But between these idyllic settings, there is a dark side: countless ships and planes have mysteriously gone missing in the one and a half million square miles of ocean separating them. (Para. 1, Text A)
2. The landscape of the island of Bermuda is quite unique: it is a remote coral reef precariously perched on a massive extinct volcano. (Para. 4, Text A)
3. Seismic activity at the bottom of the ocean can also be an explanation for disappearing ships.

Scientists have discovered that huge bubbles of methane gas can violently erupt without warning from the ocean floor and at least one oil rig is thought to have sunk because of this phenomenon. (Para. 8, Text A)

4. Astrophysical data suggests space-time might be "flat", rather than curved, and thus that it goes on forever. If so, then the region we can see (which we think of as "the universe") is just one patch in an infinitely large "quilted multiverse". (Para. 7, Text B)
5. According to the current theories, if you were to drop a cube of iron into a black hole, there would be no way to retrieve any of that information. That's because a black hole's gravity is so strong that its escape velocity is faster than light—and light is the fastest thing there is. (Para. 10, Text B)

Additional Reading

Woman's Transplanted "Man Hands" Became Lighter and More Feminine Over Time

1. A young woman in India who lost both of her hands in a bus accident received limbs from a darker-skinned male donor. Years later, the skin of her transplanted hands has lightened.
2. After her accident in 2016, 18-year-old Shreya Siddanagowder's arms were amputated below the elbow. In 2017, she underwent a 13-hour transplant operation performed by a team of 20 surgeons and 16 anesthesiologists, The Indian Express reported on March 7.
3. Her transplanted hands came from a 21-year-old man who died after a bicycle crash. Over the next year and a half, physical therapy improved Siddanagowder's motor control of her arms and hands, which gradually became leaner than they were at the time of the transplant. But there was another unexpected change: The skin on her new limbs, which had been darker because the donor had a darker complexion, became lighter in color, so that it more closely matched Siddanagowder's skin tone, according to The Indian Express.
4. The doctors who treated Siddanagowder suspect that her body produces less melanin than her donor's did, which could explain the lightening of her new limbs (melanin is a pigment that lends skin its color). But more research is required to confirm the cause, Dr. Uday Khopkar, head of dermatology at King Edward Memorial Hospital in Mumbai, told The Indian Express.
5. Candidates for hand transplants undergo evaluations and consultations that can span

months, according to the Mayo Clinic. Experts assess the patient's overall health, conducting blood tests and X-rays and evaluating nerve function in the amputated limbs. Eligible applicants are then placed on a waiting list and are matched with hand donors based on factors such as skin color, hand size and blood type, the Mayo Clinic says.

6. Siddanagowder's visit to the transplant center at the Manipal Institute of Technology in Karnataka, India, to register for a transplant coincided with a hand donation that matched her blood type. Her surgery was the first double hand transplant performed in Asia, as well as the continent's first intergender limb transplant, The Indian Express reported.
7. "I am the first female in the world to have male hands," Siddanagowder said in a video shared on Facebook in June 2019 by the MOHAN Foundation, a charitable nongovernmental organization that supports pioneering research in transplantation and organ donation in India.
8. However, her hands "have feminine features now", Siddanagowder added.

Slimmer and Lighter

9. One explanation for her hands taking on a more "feminine" shape could be the muscles adapting to their new host, physiotherapist Ketaki Doke, who worked with Siddanagowder in her home city of Pune, told The Indian Express.
10. "The nerve begins to send signals—it is called reinnervation—and the muscles function according to body needs," Doke said. "The muscles in her hand may have started adapting to a female body."
11. In the video, Siddanagowder rolled up her left sleeve to show where the transplanted forearm joined her arm, noting that its formerly darker color had lightened since she received the transplant in 2017.
12. "Now it matches my own skin color," she said.
13. Fewer than 100 people have received hand transplants worldwide, according to Johns Hopkins Medicine in Baltimore. Siddanagowder's doctors are monitoring the changes in her hands' skin color and shape, and they expect to publish the details of her transplant and recovery in a case report, according to Dr. Subramania Iyer, head of plastic and reconstructive surgery at the Amrita Institute of Medical Sciences in Kerala, India.
14. However, more evidence will be required to understand what is driving these changes in her transplanted hands, Iyer told The Indian Express.